Engelbert Redel

Transition Metal Nanoparticles

Engelbert Redel

Transition Metal Nanoparticles

Synthesis, Characterization, Stabilization and Functionalization of different Transition Metal Nanoparticles in Ionic Liquids (ILs).

Südwestdeutscher Verlag für Hochschulschriften

Impressum/Imprint (nur für Deutschland/ only for Germany)
Bibliografische Information der Deutschen Nationalbibliothek: Die Deutsche Nationalbibliothek verzeichnet diese Publikation in der Deutschen Nationalbibliografie; detaillierte bibliografische Daten sind im Internet über http://dnb.d-nb.de abrufbar.
Alle in diesem Buch genannten Marken und Produktnamen unterliegen warenzeichen-, marken- oder patentrechtlichem Schutz bzw. sind Warenzeichen oder eingetragene Warenzeichen der jeweiligen Inhaber. Die Wiedergabe von Marken, Produktnamen, Gebrauchsnamen, Handelsnamen, Warenbezeichnungen u.s.w. in diesem Werk berechtigt auch ohne besondere Kennzeichnung nicht zu der Annahme, dass solche Namen im Sinne der Warenzeichen- und Markenschutzgesetzgebung als frei zu betrachten wären und daher von jedermann benutzt werden dürften.

Verlag: Südwestdeutscher Verlag für Hochschulschriften Aktiengesellschaft & Co. KG
Dudweiler Landstr. 99, 66123 Saarbrücken, Deutschland
Telefon +49 681 37 20 271-1, Telefax +49 681 37 20 271-0, Email: info@svh-verlag.de
Zugl.: Freiburg (in Breisgau), Albert-Ludwigs-University Freiburg, Diss., 2009

Herstellung in Deutschland:
Schaltungsdienst Lange o.H.G., Zehrensdorfer Str. 11, D-12277 Berlin
Books on Demand GmbH, Gutenbergring 53, D-22848 Norderstedt
Reha GmbH, Dudweiler Landstr. 99, D- 66123 Saarbrücken
ISBN: 978-3-8381-1026-4

Imprint (only for USA, GB)
Bibliographic information published by the Deutsche Nationalbibliothek: The Deutsche Nationalbibliothek lists this publication in the Deutsche Nationalbibliografie; detailed bibliographic data are available in the Internet at http://dnb.d-nb.de.
Any brand names and product names mentioned in this book are subject to trademark, brand or patent protection and are trademarks or registered trademarks of their respective holders. The use of brand names, product names, common names, trade names, product descriptions etc. even without
a particular marking in this works is in no way to be construed to mean that such names may be regarded as unrestricted in respect of trademark and brand protection legislation and could thus be used by anyone.

Publisher:
Südwestdeutscher Verlag für Hochschulschriften Aktiengesellschaft & Co. KG
Dudweiler Landstr. 99, 66123 Saarbrücken, Germany
Phone +49 681 37 20 271-1, Fax +49 681 37 20 271-0, Email: info@svh-verlag.de

Copyright © 2008 Südwestdeutscher Verlag für Hochschulschriften Aktiengesellschaft & Co. KG and licensors
All rights reserved. Saarbrücken 2008

Produced in USA and UK by:
Lightning Source Inc., 1246 Heil Quaker Blvd., La Vergne, TN 37086, USA
Lightning Source UK Ltd., Chapter House, Pitfield, Kiln Farm, Milton Keynes, MK11 3LW, GB
BookSurge, 7290 B. Investment Drive, North Charleston, SC 29418, USA
ISBN: 978-3-8381-1026-4

- Transition Metal Nanoparticles -

Preface:

This thesis describes the reproducible synthesis of different transition metal nanoparticles (M-NPs with M = Cr, Mo, W, Fe, Ru, Os, Co, Rh, Ir, Ag and Au) from different precursors (metal salts, metal complexes and organometallic compounds) using different synthetic methods (H_2 reduction, in situ salt reduction, thermolytic, photolytic and microwave based decomposition and reduction processes) in ionic liquids (ILs). The ionic liquids act as non-surfactant or weakly coordinating supramolecular network for the kinetic stabilization of the metal nanoparticles. The very small and uniform nanoparticle size of about 1 to 3 nm in $BMIm^+BF_4^-$ increases with the molecular volume of the ionic liquid anion in $BMIm^+PF_6^-$, $BMIm^+OTf^-$ and $BtMA^+NTf_2^-$ ($BMIm^+$ = n-butyl-methyl-imidazolium, $BtMA^+$ = n-butyl-trimethyl-ammonium, NTf_2 = $N(O_2SCF_3)_2$, OTf = O_3SCF_3). Characterization of the dispersed transitions nanoparticles was done by transmission electron microscopy (TEM and HRTEM, size determination), transmission electron diffraction (TED, crystallinity and verification of metal state versus metal oxide identity), X-ray powder diffraction (XRPD, crystallinity, metal state identity and size approximation from Scherrer equation) and dynamic light scattering (DLS, size determination, hydrodynamic radius). Under argon the M-NP/IL dispersions are kinetically stable without any additional stabilizers or capping molecules.

We suggest that ionic liquids act as a "*novel nanosynthetic template*", no extra stabilizers or capping molecules are needed. Stable Ag nanoparticles are obtained reproducibly by H_2 reduction of different Ag(I)X salts (X = BF_4, PF_6, OTf) dissolved ILs in the presence of n-butyl-imidazole (Bim) as a scavenger for the HX acid byproduct in order to avoid disturbance of the IL network and, thus, Ag-NP destabilization.

Metal carbonyls were introduced as precursors for M-NP synthesis in ILs to avoid the problems during H_2 reduction of metal salts: Stable Cr-, Mo-, W-, Fe-, Ru-, Os-, Co-, Rh- and Ir-NPs are obtained reproducibly by thermal or photolytic decomposition under argon from the metal carbonyl precursors $M(CO)_6$ (M = Cr, Mo, W), $Fe_2(CO)_9$, $Ru_3(CO)_{12}$ or $Os_3(CO)_{12}$, $Co_2(CO)_8$, $Rh_6(CO)_{16}$ and $Ir_4(CO)_{12}$, respectively, suspended in the ionic liquids $BMIm^+BF_4^-$, $BMIm^+OTf^-$ and $BtMA^+NTf_2^-$. The Ru-, Rh- and Ir-NP/IL systems function as highly effective and recyclable catalysts in the biphasic liquid-liquid hydrogenation of cyclohexene to cyclohexane.

Gold nanoparticles (Au-NPs) were prepared by thermal decomposition and reduction of Au(CO)Cl or $KAuCl_4$ in the presence of n-butyl-imidazole as scavenger in different ILs. The Au-NPs can then be transferred to polar and non-polar organic solvents by introducing organic capping molecules like thiolglycolic acid and n-decanethiole or onto a polytetrafluoroethylene (PTFE) surface. Attempts to deposit the Au-NPs onto carbon nanotubes failed.

The surrounding and stabilization of Au-NPs with the IL anions is supported by DFT calculations and was found to be much stronger than the interaction to imidazolium cations. The sufficiently but not too strongly stabilizing IL network properties towards M-NPs allowed us to succeed in a step by step sequential growth of Au-NPs in ionic liquids, which was monitored by the specific colour changes and an increase in the Au-SPR (Surface Plasmon Resonance) band for metallic Au nanoparticles during nucleation and growth.

Engelbert Redel

Freiburg (im Breisgau)
May 2009

This work is dedicated to *Richard Zsigmondy* (Nobel Laureate) 1865-1929.

— Nobel Prize in Chemistry 1925 —

– In remember of his glorious work about gold-colloids –

"Für die Aufklärung der heterogenen Natur kolloidaler Lösungen sowie für die dabei angewandten Methoden, die grundlegend für die moderne Kolloidchemie sind"

In the Beginning
There was Nano ...

(from G. A. Ozin "The WIZARD of OZ", Nanochemistry, 2nd. Ed., RSC 2009)

Acknowledgements

Firstly I would like to thank Prof. Dr. Christoph Janiak for his generosity, his great support, especially in publishing, and his guidance to my doctoral research. This big freedom which he offered me was at least the main key for our success. More importantly, I would like to thank him for his friendship, on both scientific and personal levels, that he has provided throughout the past three years. Through his guidance, I was able to approach my own research, in a way that has widened my knowledge and perspective on *Nanomaterials* and *Nanochemistry* tremendously.

Secondly, I thank Prof. Dr. Margit Zacharias for the co reference of my Ph.D. work. I want to praise her, for those great lectures about the fascinating topic of *Nanotechnology* and *Nanomaterials*. These lectures open my eyes to a new dimension of understanding nature and science.

I also thank Prof. Dr. Gerald Urban for the co reference of my Ph.D. work. I want to thank him also, for his kind manner and his helpful suggestions and discussion about our common project and different application in the field of *Nanomaterials* and *Nanotechnology*.

Big thanks go to Dr. Michael Krueger for the very nice time which he offered me in his Nanolab at FMF (Freiburger-Material-Forschungszentrum). This time was an important mark in my Ph.D. time; I learnt here, how a successful collaboration really works.

A special thank goes to Dr. Ralf Thomann for the countless TEM measurements which he has done for me with great motivation and ambition. Also the common discussion and his advice were very useful for our various common nanotechnology publications.

A great thank goes to Dr. Michael Walter for his interest in ionic liquids (ILs) and his ambition and motivation for DFT calculations in the field of ILs and Au-Nanoclusters. Also I want to thank Dr. H. Scherer for his help and interest in the dynamic NMR measurement of dispersed Au-Nanoclusters in the dynamic IL-Matrix.

The following persons were indispensable to the progress of my doctoral work and the field of ionic liquids. Firstly, I want to thank Dr. Thomas Schubert and Dr. Tom Beyersdorf from IoLiTec, for their interest and support of my research work. Secondly, I also want to thank Dipl.-Chem. Jérôme Krämer and Dipl.-Chem. Christian Vollmer for their interest in ionic liquids and nanocatalysis, also for the continuation of this research area.

Of the other collaborators in the FMF Nanolab, a special thank goes to Frank Riehle, "*the unbreakable*", for the funny time which we had together, that I will never forget. Also a great thank you is for Laith Hussein and Dr. Gregorgy B. Stevens, two fine and very helpful persons, who supported me in my research, and create a pleasant atmosphere in Krueger´s Nanolab. I also want to thank Ying Yaun, Yunfei Zhou and Ruonan Zhai for a good time during my stay and a really nice group atmosphere.

A great thank goes also to Prof. Dr. Caroline Röhr and Marco Wendorff for their help, measurements and calculation of compound $\{(C_4H_{12}N_2)_2[Cu^I I_4](I_2)\}_n$, which can be regarded as the first inorganic model for the classical starch-iodine compound. Furthermore, I want to thank Boumahdi Benkmil and Dr. Teame Tekeste from the workgroup of em. Prof. Dr. H. Vahrenkamp, for useful discussions and the nice time we had together.

To all other collaborators in the Janiak workgroup and former members, like Frederik Blank, Anne Chamayou, Reda El-Shaarawy, Jana Vieth, Viktoria Gräfner, Hesham Mena, Barbara Wisser, André Paske, Khalid Abu Shandi, Angela Thiemann and Sabine Zuelsdorf, I want to thank them all for a nice time during the last three years, some funny situations, interesting scientific and non-scientific discussions and a pleasant workgroup atmosphere.

Lastly I thank the people who are the most important to me – my family and my parents, Georg and Hilde Redel – for the years of care and sacrifice they have made to provide the education and life I am privileged to have. I also thank their open-mindedness in allowing me to discover my own path, and their continuous belief in my ability and scientific potential.

The following parts of this thesis are already published, are in press or have been submitted for publication.

Patent Applications:

C. JANIAK, E. REDEL, M. KLINGELE, T. F. BEYERSDORF, T. J. S. SCHUBERT,
Verfahren zur Herstellung von metallhaltigen Nanopartikeln; (Univ. Freiburg und IOLITEC)
DE-Patentanmeldung, **DE 10 2007 045 878.0-24** (25. 09. 2007); PCT/EP-Patentanmedlung, **PCT/EP2008/008084** (30.09.2008); **WO2009/040107** (04.02.2009).

C. JANIAK, E. REDEL, M. KLINGELE, T. F. BEYERSDORF, T. J. S. SCHUBERT,
Verfahren zur Herstellung und Stabilisierung von funktionellen Metallnanopartikeln in ionischen Flüssigkeiten; (Univ. Freiburg und Firma IOLITEC).
DE-Patentanmeldung, **DE 10 2007 038 879.0** (17. 08. 2007) PCT/EP-Patentanmeldung, **PCT/EP2008/006768** (23.10.2008); **WO2009/024312** (26.02.2009).

Publications:

E. REDEL, M. WALTER, R. THOMANN, L. HUSSEIN, M. KRÜGER, C. JANIAK
Step by step of controlled and sequential growth of Au clusters and nanoparticles in Ils.
submitted, **2009** (for Nature Nanotechnology)

C. VOLLMER, E. REDEL, K. ABU-SHANDI, R. THOMANN, C. JANIAK
Use of ionic liquids for the synthesis of different transition metal nanoparticles under mild microwave and photolytic conditions from their metal carbonyl precursors and the use of Ru-NP/IL as bisphasic liquid-liquid hydrogenation nanocatalysts for cyclohexene.
submitted, **2009** (for Chem. Eur. J.)

E. REDEL, M. WALTER, R. THOMANN, C. VOLLMER, L. HUSSEIN, H. SCHERER, M. KRÜGER, C. JANIAK
Synthesis, stabilization, functionalization and DFT calculations of gold nanoparticles in fluorous phases (PTFE and ILs);
Chem. Eur. J. **2009**, in press; DOI 10.1002/chem.200900301

E. REDEL, J. KRÄMER, R. THOMANN, C. JANIAK
Synthesis of Co, Rh and Ir nanoparticles from metal carbonyls in ionic liquids and their use as bisphasic liquid-liquid hydrogenation nanocatalysts for cyclohexene.
(Dedicated to Prof. Christoph Elschebroich on the occasion of his 70th birthday)
J. Organomet. Chem. (Special Issue) **2009**, *694*, 1069-1075.

J. KRÄMER, E. REDEL, R. THOMANN, C. JANIAK
Use of ionic liquids for the synthesis of Fe, Ru and Os nanoparticles from their metal carbonyl Precursors.
Organometallics **2008**, *27*, 1976-1978.

E. REDEL, J. KRÄMER, R. THOMANN, C. JANIAK
Ionische Flüssigkeiten als Templat für Nanosynthesen; Ionic liquids as template for nanosynthesis. Synthesis of customized metal containing functional nanoparticles.
GIT Labor Fachzeitschrift **2008**, *52*, 400-403.

E. REDEL, R. THOMANN, C. JANIAK
The use of ionic liquids (ILs) for the IL-anion size-dependent formation of Cr, Mo and W nanoparticles from metal carbonyl $M(CO)_6$ precursors.
Chem. Commun. **2008**, *15*, 1789-1791.

E. REDEL, R. THOMANN, C. JANIAK
First correlation of nanoparticle size-dependent formation with the ionic liquid anion molecular volume.
Inorg. Chem. **2008**, *47*, 14-16.

E. REDEL, M. FIEDERLE, C. JANIAK
Piperazinium, ethylenediammonium or 4,4'-bipyridinium halocuprates(I) by Cu(II)/Cu(0) comproportionation;
(Dedicated to Prof. Wilhelm Preetz on the occasion of his 75th birthday)
Z. Anorg. Allg. Chem. (Special Issue) **2009**, in press

E. REDEL, C. RÖHR, C. JANIAK
An inorganic starch-iodine model: The inorganic-organic hybrid compound $\{(C_4H_{12}N_2)_2[CuI_4](I_2)\}_n$;
Chem. Commun. **2009**, *16*, 2103-2105. (*Hot Article*, Journal Inside Cover)

E. REDEL, C. JANIAK
Speziationsanalytik von Eisenverbindungen: Polarographische Eisen(II)-/Eisen(III)-Speziesanalyse in gemischtvalenten Eisenphosphaten; Polarographic Determination of Iron(II) and Iron(III) mix-valent species in ironphosphate compounds
GIT Labor-Fachzeitschrift **2007**, *51* (5), 397-399.

E. REDEL, S. ZUELSDORF, C. JANIAK
Quantitativer Nitrit-Nachweis in Mikroreaktoren mittels Fließinjektionsanalytik; Quantitative Nitrite-Determination in Microreactors with Flow-Injection-Analysis (FIA).
in preparation, **2009**

– Table of Contents: Transition Metal Nanoparticles –

Chapter 1.1: *Introduction & Industrial Synthesis of Nanoparticles*

    1.1.1 Definition of Nanoparticles (NPs)   pp. 1

    1.1.2 Industrial Synthesis of Nanoparticles & Nanomaterials   pp. 4

    1.1.3 Lab Synthesis of Transition Metal Nanoparticles in ILs   pp.11

    1.1.4 Gold Nanoparticles (Au-NPs) and Historic Background   pp.13

Chapter 1.2: *Industrial Application of Ionic Liquids & Network properties*

    1.2.1 Introduction & Applications of ILs in the Chemical Industry   pp.15

    1.2.2 ILs Involved in Important Large Scale Industrial Processes   pp.17

    1.2.3 Network Properties of Ionic Liquids (ILs)   pp.23

Chapter 2: *Assignment*   pp.30

Chapter 3: *Results and Discussion*   pp.31

Chapter 3.1: *The First Correlation of Nanoparticle Size Dependent Formation with the Ionic Liquid Anion Molecular Volume*   pp.33

Chapter 3.2: *Use of Ionic Liquids (ILs) for the IL-Anion Size-Dependent Formation of Cr, Mo and W Nanoparticles from Metal Carbonyl $M(CO)_6$ Precursors*   pp.37

Chapter 3.3: *Use of Ionic Liquids (ILs) for the Synthesis of Fe, Ru and Os Nanoparticles from their Metal Carbonyl Precursors*   pp.42

Chapter 3.4: *Synthesis of Co, Rh and Ir Nanoparticles from Metal Carbonyls in ILs and their Use as Biphasic Liquid-Liquid Hydrogenation Nanocatalysts for Cyclohexene*   pp.47

Chapter 3.5: *Synthesis, Stabilization, Functionalization and DFT Calculations of Gold Nanoparticles in Fluorous Phases (PTFE and ILs)*

    3.5.1 Au-NP synthesis   pp.54

    3.5.2 Au-NP surface functionalization   pp.59

    3.5.3 Au-NP deposition on PTFE (Teflon)   pp.63

    3.5.4 Dynamic NMR studies   pp.66

    3.5.5 DFT calculations   pp.67

Chapter 3.6: *Stepwise, Ligand-Free and Controlled Growth of Gold Nanoparticles in Ionic Liquids (ILs)* pp.69

Chapter 4: *Conclusions* pp.74

Chapter 5: *Instruments & Experimental Section*
    5.1 Instrumentation and Devices pp.76
    5.2 Synthesis equipment pp.78
    5.3 Synthesis procedure and details pp.80

References pp.86

Appendix: Organic-Inorganic Hybrid Materials

Chapter 1.1: Introduction & Industrial Synthesis of Nanoparticles

1.1 Introduction:

Nanoparticles (NPs) are ultrafine particles with sizes of the order of a nanometer. *"Nano"* is a prefix denoting ten to the power of minus 9, namely one billionth. One nm is an extremely small length, corresponding to one billionth of 1 m, one millionth of 1 mm, or one thousandth of 1 μm. The definition of *"nanoparticles"* differs depending upon the materials, fields and applications concerned. Note that at small size, nanoparticles have a high percentage of surface atoms. (Fig. 1.1b and Table 1.1) In the narrower sense, nanoparticles are regarded as particles smaller than 10–20 nm, where the physical properties of solid materials themselves would drastically change.[1]

1.1.1 Definition of Nanoparticles (NPs):

Nanomaterials and Nanoparticles, characterized by at least one dimension in the nanometer range, can be considered to constitute a bridge between single molecules and infinite bulk systems (see Fig.1.1.1).[2] Suitable control of the properties of nanometer-scale structures can lead to new science as well as new devices and technologies. Besides individual nanostructures involving clusters, nanoparticles, quantum dots, nanowires and nanotubes, collections of these nanostructures in the form of arrays and superlattices are of vital interest to the science and technology of nanomaterials.[3] Chemistry, especially inorganic chemistry plays a particularly important role in the synthesis and characterization of nano-building units such as nanocrystals of metals, oxides and semiconductors, nanoparticles and composites involving ceramics, nanotubes of carbon and inorganics, nanowires of various materials and polymers.[4] Assembling these units into arrays also involves chemistry. Although the area of nanoscience is young, it seems likely that new devices and technologies will emerge in the near future. Electrochemistry and photochemistry using nanoparticles and nanowires, and nanocatalysis of different nanocluster or nano-particles are examples of such new chemistry.[5] Moreover, nanoporous solids have been attracting increasing attention in the last few years.[6]

The importance of nanotechnology was pointed out by Feynman as early as 1959, in his often-cited lecture entitled *"**There is plenty of room at the bottom**"*.[7] The structure and properties of nanomaterials differ significantly from those of atoms and molecules as well as those of bulk materials. Synthesis, structure, energetics, response, dynamics and a variety of other properties and related applications form the theme of the emerging area of nanoscience, and there is a large chemical component in all of these aspects.[8]

Size effects constitute a fascinating aspect of nanomaterials.[9] The effects determined by size pertain to the evolution of structural, thermodynamic, electronic, spectroscopic, electro-magnetic and chemical features of these finite systems with increasing size. Size effects can be classified into two types, one dealing with specific size effects (e.g. magic numbers of atoms in metal clusters, quantum mechanical effects at small sizes) and the other involving size-scaling applicable to relatively larger nanostructures. The former includes the appearance of new features in the electronic structure. Figure 1.1.1a show how the electronic structures of metal or semiconductor nanocrystals differ from those of bulk materials and isolated atoms.

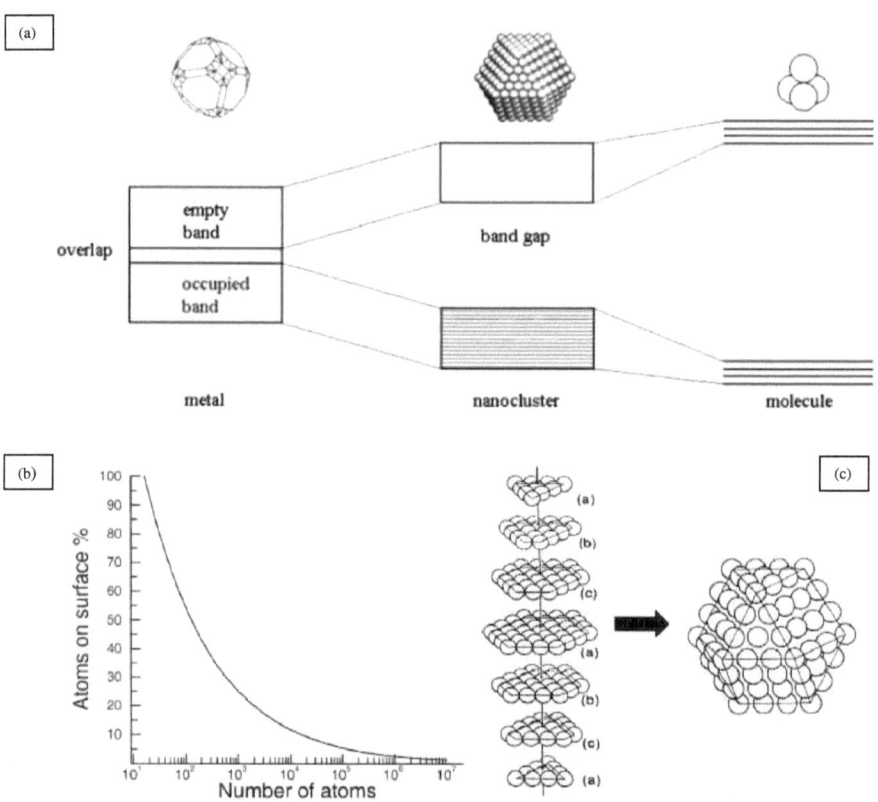

Fig. 1.1.1: (a) Formation of discrete electronic energy levels on the way from bulk to molecule (figure from reference) (b) Plot of the number of atoms (*n*) vs. the percentage of atoms located on the surface of a particle, valid for metal particles (figure from reference). (c) Schematic illustration of how a cuboctahedral 147 atom-cluster, composed of seven close-packed layers can be made out of a stacking sequence reminiscent of a fcc lattice (figure from reference).[10]

Table 1.1: Diameter (Ø nm) of metallic Cu and Au nanoparticles as a function of n (number of atoms).

n	Ø Cu (nm)	Ø Au (nm)
10^1	0.6	0.7
10^2	1.3	1.5
10^3	2.8	3.2
10^4	6.1	6.8
10^5	13.1	14.7
10^6	28.2	31.8
10^7	60.9	68.5
10^8	131.1	147.5

However, an interesting interplay exists between the morphology and the packing arrangement, especially in small nanoclusters and nanocrystals. If one were to assume that metal nanoparticles strictly follow the bulk crystalline order, the most stable structure is arrived at by simply constraining the number of surface atoms.[11] (Fig. 1.1.1c) Moreover, the properties of nanoparticles are also influenced by the formation of geometric shells which occur at definite nuclearities.[12] Such nuclearities (N), called *magic nuclearities* endow a special stability to nanoparticles as can be demonstrated on the basis of purely geometric arguments. A new shell (L) of a particle emerges when the coordination sphere of an inner central atom or shell (forming the previous shell) is completely satisfied. The number of atoms or spheres required to complete successive coordination shells is a problem that mathematicians, starting with Kepler, have grappled with for a long time.[13] (Fig. 1.1.2)

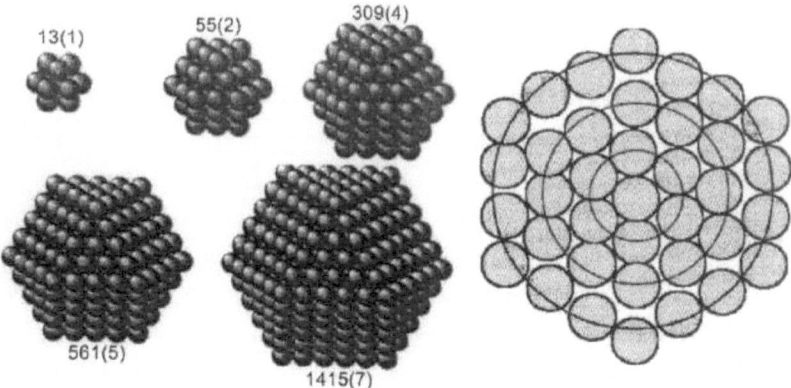

Fig. 1.1.2: Nanoparticles in closed-shell configurations with magic number of atoms. The numbers beside correspond to the nuclearity (N) and the number of shells (L). The figure on the left is a cross-sectional view showing five coordination shells in a 561 atom cluster (figure from reference).[13]

The field of Nanochemistry[14] & Nanomaterials has matured so rapidly and so fast that it is probably hard to find a segment of any technical subject where the implications of nano-materials have not been explored at least to a preliminary extent. Studies are being conducted on the potential use of nanomaterials in diverse applications, including hydrogen storage,[15,16] ion-sensing and gas sensing,[17] surface-modified nanoparticles for enhanced oil recovery,[18] adsorption of chemical and biological agents on to nanoparticles,[19] active electrode materials for lithium-ion batteries,[20] light-emitting devices[21] and dental compositions, to mention only of few. Over the past 5-7 years, nanoparticle producers have been working hard to differentiating themselves from their competitors, by both providing nanopowders and nanomaterials with varied particle characteristics or by developing nanoparticles of proprietary compositions. In many instances, the ability to supply tonnage quantities of nanopowders has also been established. The selling price (which in many cases is a function of the manufacturing cost) has also come down over the years. Especially applications where only a small amount, say approximately 1 to 10 wt%, of nanoparticle addition has been able to change substantially the properties and performance of the end-product are becoming increasingly popular. A number of such examples can be found in the area of functional coatings.

1.1.2 Industrial Synthesis of Nanoparticles & Nanomaterials

Nanostructured materials, particularly those derived from nanoparticles, have evolved as a se-parate class of materials over the past decade. The most remarkable feature has been the way in which completely disparate disciplines have come together with nanomaterials as a theme.

All industrial particle synthesis techniques fall into one of the four categories: *vapor-phase, sputtering processes, solution precipitation, and solid-state processes.* There are a handful of processes that combine aspects of one or more of these broad categories of processes. Although vapor-phase processes have been in vogue during the early days of nanoparticles development, the solid-state process is the most widely used in the industry for production of micron-sized particles, predominantly due to cost considerations. One of the most established powder producers, Ferro Corporation, uses the *solid-state synthesis* method almost exclusively; hundreds of tons of lithium cobalt oxide, which is commonly used as cathode material in lithium-ion batteries, is produced using solid-state synthesis.

1.1.2.1 SOLID-STATE SYNTHESIS OF NANOPARTICLES

Solid-state synthesis generally involves a heat treatment step (in order to achieve the desired crystal structure), which is followed by media milling.[22] While it is generally believed that it is difficult for the lower limit of the average particle size to be much below 100 nm, recent innovations by established companies in the industry may prove otherwise. Judging, by the contents of publications, the scientific community has not shown much enthusiasm for mechanical attrition processes for nanoparticles synthesis, perhaps due to issues pertaining to impurity pick up, lack of control on the particle size distribution, and inability to tailor precisely the shape and size of particles in the 10 to 30 nm range, as well as the surface characteristics.[23] Nonetheless, in several instances a modified version of mechanical attrition has been used to synthesize oxide nanoparticles.

1.1.2.2 VAPOR-PHASE SYNTHESIS OF NANOPARTICLES

Gas condensation, as a technique for producing nanoparticles, refers to the formation of nanoparticles in the gas phase, i.e., condensing atoms and molecules in the vapor phase. Cabot Corporation in the United States and Degussa/Evonik in Germany, have been using atmospheric flame reactors for decades to produce megatons of such diverse nanoparticles as carbon black (used in tires and inks), silicon dioxide (used in myriad applications including additives in coffee creamers and polymers), and titanium dioxide (used in scores of applications including UV-protecting gels).[24]

Flame-Based Synthesis of Nanoparticles (PVS and CVC)

The use of a hydrocarbon (or hydrogen)–oxygen flame to pyrolyze chemical precursor species and produce nanoparticles is attractive in principle due to the fact that flame processes are already in use on a commercial scale. Over the past decade and a half, research has been directed predominantly toward introducing uniformity and control over the pyrolysis process in a flame, with the anticipation of forming nanoparticles with a narrow size distribution and minimal aggregation. This included developing flames with a flat geometry, as opposed to the traditional Bunsen burner conical flames.

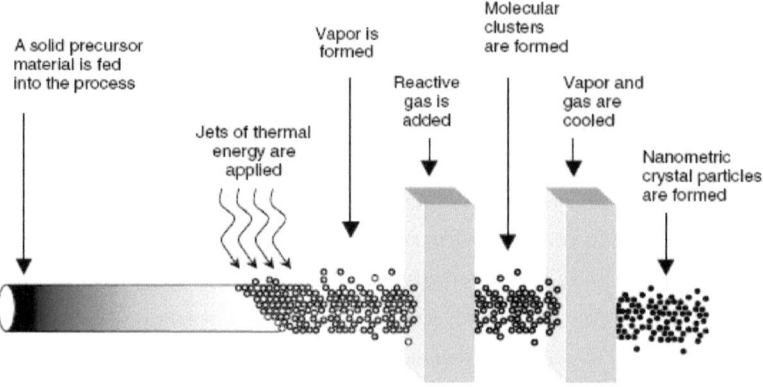

Fig. 1.1.3: Physical vapor synthesis (PVS) method for the synthesis of NPs (figure from reference).[25]

The synthesis of nanoparticles by the physical vapor synthesis (PVS) method, first introduced by Granqvist and Buhrmann,[26] and Gleiter et al.[27] (Fig. 1.1.3), involves evaporation of metal species under a reduced pressure of inert gas in a ultrahigh-vacuum chamber. Nanoparticles develop in a thermalizing zone just above the evaporative source due to interactions between the hot vapor species and the much colder inert gas atoms in the chamber. In this process (Fig. 1.1.3), precursor material is introduced at a controlled rate into a chamber. In the chamber, a plasma arc is formed between a nonconsumable electrode and the precursor. The precursor, typically a high-purity metal rod, passes through the plasma arc and is melted and vaporized. A reactive gas, typically oxygen, is introduced into the chamber and reacts with the eva-porated precursor, causing nanoparticles of oxides to be formed upon condensation. Additional gas is turbulently introduced to accelerate cooling of the particles. Nanopowders of Fe_2O_3, Al_2O_3, TiO_2, and CeO_2 are routinely produced in tonnage quantities by the PVS method under the trade name Nanotek.

The synthesis of nanoparticles by the chemical vapor condensation (CVC) method, first intro-duced by Chang et al.,[28] involves controlled thermal decomposition of organometallic pre-cursors in a reduced-pressure environment. By using a hot-wall reactor and an inert carrier gas for the precursor, nonoxide ceramic nanopowders can be synthesized. Using a combustion flame reactor allows oxide ceramics to be produced (Fig. 1.1.4). The CVC process utilizes pre-cursor vapors, and is restricted to oxide ceramics that could be derived from metalorganic or organometallic precursors with ambient pressure and boiling points of ~200°C or lower.

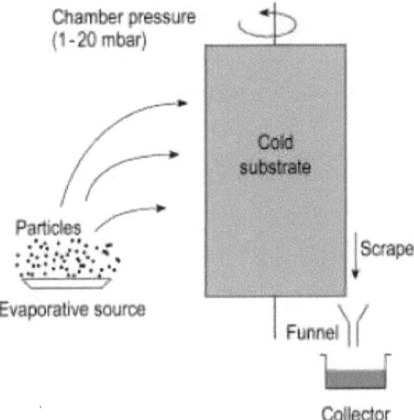

Fig. 1.1.4: Chem. vapor condensation (CVC) method for the synthesis of NPs (fig. from reference).[29]

The PVS process involves evaporation of elemental species, such as Al and Fe, which are subsequently oxidized in the gas phase, while the CVC processes involve pyrolysis of vapors of organometallic compounds and halides, respectively, in a hot zone, followed by condensation.[30]

Spray Pyrolysis of Nanoparticles

Spray pyrolysis, which combines aspects of gas-phase processing and solution precipitation, has been in use for quite some time. A company by the name Seattle Specialty Ceramics, employed spray pyrolysis to produce specialty powders. The technique, shown schematically in Fig. 1.1.5, involves the formation of precursor aerosol droplets that are delivered by a carrier gas through a heating zone.[31]

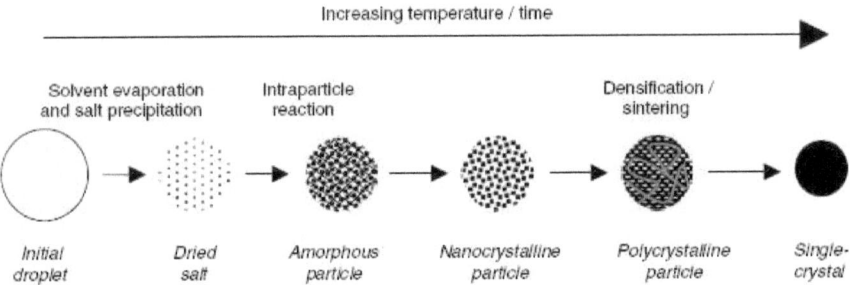

Fig. 1.1.5: Sequence of events during the spray pyrolysis process (figure from reference).[32]

Precursor solutions of metal nitrates, metal chlorides, and metal acetates are atomized into fine droplets and sprayed into the thermal zone. Inside the heating zone, the solvent eva-porates and reactions occur within each particle to form a product particle. Spherical, dense particles in the 100 to 1000 nm range can easily be formed in large volume by this method. The principal advantage of the spray pyrolysis method is the ability to form multicomponent nanoparticles as solutions of different metal salts can be mixed and aerosolized into the reaction zone. Over the years, much emphasis has been given to being able to reduce the precursor droplet size during spray pyrolysis as this in turn would reduce the particle size. Tsai et al.[33] pushed the limit on the particle size by using precursor drops that were 6 to 9 μm in diameter.

1.1.2.3 SPUTTERING PROCESSES

In this method, accelerated ions such as Ar^+ are directed toward the surface of a target to eject atoms and small clusters from its surface. The ions are carried to the substrate under a relatively high pressure (~1mTorr) of an inert gas, causing aggregation of the species. Nanoparticles of metals and alloys as well as semiconductors have been prepared using this method. Urban et al.[34] have demonstrated the formation of nanoparticles of various metals using magnetron sputtering. They formed collimated beams of nanoparticles and deposited them as nanostructured films on Si substrates. Birtcher et al. have made an interesting observation of Au nanoparticle formation from single Xe ion impacts.[35]

1.1.2.4 SOLUTION PROCESSING OF NANOPARTICLES

Precipitating clusters of inorganic compounds from a solution of chemical compounds has been an attractive proposition for researchers, primarily because of the simplicity with which experiments can be conducted in a laboratory. This is especially true if the goal is to just have a nanocrystalline powder, instead of a "dispersible" nanoparticulate powder.[36]

Solution processing can be classified into five major categories:
(1) sol–gel processing, (2) solution precipitation method, (3) water–oil microemulsions (reverse micelle) method, (4) polyol method, and (5) hydrothermal synthesis.

A major advantage of solution processing is the ability to form encapsulated nanoparticles, specifically with an organic molecule, for providing functionality to the nanoparticles, im-proving their stability in a medium, or for controlling their shape and size.[37]

(1) *Sol-gel Processing*

The Sol–gel technique is one of the most popular solutions processing method for producing metal oxide nanoparticles. This process is well described in several books[38,39] and reviews. Over the years, solution precipitation and sol–gel processing have come to be used interchangeably, mostly by people on the fringes of the technical community. There are distinct differences between the two methods, as will be made clear below. In sol–gel processing, a reactive metal precursor, such as metal alkoxide, is hydrolyzed with water, and the hydrolyzed species are allowed to condense with each other to form precipitates of metal oxide nanoparticles. The precipitate is subsequently washed and dried, which is then calcined at an elevated tem-perature to form crystalline metal oxide nanoparticles.

The hydrolysis of metal alkoxides involves nucelophilic reaction with water, which is as follows:

$$M(OR)_y + x\,H_2O \rightarrow M(OR)_{y-x}(OH)_x + x\,ROH$$

Condensation occurs when either hydrolyzed species react with each other and release a water molecule, or a hydrolyzed species reacts with an unhydrolyzed species and releases an alcohol molecule. The rates at which hydrolysis and condensation reactions take place are important parameters that affect the properties of the final product.

(2) *Solution Precipitation Method*

In the precipitation method, an inorganic metal salt (e.g., chloride, nitrate, acetate, or oxychloride) is dissolved in water. Metal cations in water exist in the form of metal hydrate species, such as $Al(H_2O)^{3+}$ and $Fe(H_2O)_6^{3+}$. These species are hydrolyzed by adding a base solution, such as NaOH or NH_4OH. The hydrolyzed species condense with each other to form either a metal hydroxide or hydrous metal oxide precipitate on increasing the concentration of OH^- ions in the solution. The precipitate is then washed, filtered, and dried. The dried powder is subsequently calcined to obtain the final crystalline metal oxide phase. The major advantage of this process is that it is relatively economical and is used to synthesize a wide range of single-and multicomponents oxide nanopowders.[40]

(3) *Water–Oil Microemulsiom (Reverse Micelle) Method*

Uniform and size-controlled nanoparticles of metal, semiconductor, and metal oxides can be produced by the water-in-oil (W/O) microemulsion (also called reverse micelle) method. In a W/O microemulsion, nanosized water droplets, stabilized by a surfactant, are dispersed in an oil phase.[41] A schematic view of a W/O microemlusion is shown in Fig. 1.1.6.

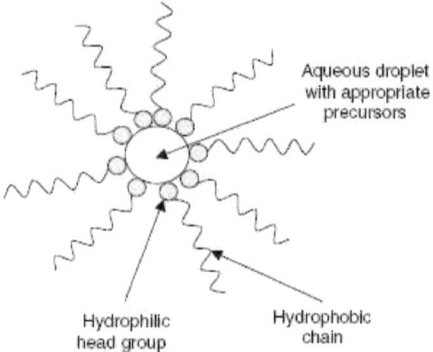

Fig. 1.1.6: Schematic of a reverse micelle.[42]

Nanosized water droplets act as a *microreactor*, wherein particle formation occurs and helps to control the size of nanoparticles. A unique feature of the reverse micelle process is that the particles are generally nanosized and monodisperse[43]. This is because the surfactant mole-cules that stabilized the water droplets also adsorb on the surface of the nanoparticles, once the particle size approaches that of the water droplet.

A common way to practice the reverse micelle technique is by mixing two microemulsions that carry appropriate reactants. Water droplets of two microemlusions are allowed to collide with each other and the particle formation reaction takes place inside the water droplet. Nanoparticle synthesis inside reverse micelles is accomplished by one of the two different chemical reactions: (1) hydrolysis of metal alkoxides or precipitation of metal salts with a base, in case of metal oxide nanoparticles, and (2) reduction of metal salts with a reducing agent, such as $NaBH_4$, in case of metal nanoparticles. Particles are either filtered or centri-fuged and then washed with acetone and water to remove any residual oil and surfactant molecules adsorbed on the surface of nanoparticles[44]. Subsequently, the powders are calcined to form the final product.

(4) Polyol Synthesis Method

The Polyol synthesis method is a simple method for forming metal nanoparticles, wherein metal acetates or other suitable salts or organometallic precursors are thermally decomposed in the presence of a passivating solvent, such as glycol ether. When the salt/precursor–solvent mixture is refluxed for a prolonged period of time at a temperature above the melting point of the salt/precursor, metal ions come together to form nanoparticles. It will be believe that the solvent binds to the surface of the metal clusters, thereby retarding growth and aggregation into larger particles. As expected, the nanoparticle characteristics are dependent on the concentration of the salt or organometallic precursors in the solvent.[45]

(5) Hydrothermal Synthesis Method

The hydrothermal method, deals with the different aspects of solution precipitation techniques. Nanocrystalline oxide ceramic particles in the range 100 nm to a few microns have been produced by hydrothermal processing.[46] The method of producing oxide nanoparticles by the hydrothermal technique begins with the co-precipitation of the metal oxide components. The precipitated slurry is then drawn into a device where it is immediately elevated to high pressures and temperature within a small volume. The precipitated gel experiences ultrashear forces and cavitational heating during the synthesis. These two aspects lead to the formation of nanophase particles and high-phase purity of complex metal oxides.[47]

1.1.3 Lab Synthesis of Transition Metal Nanoparticles in Ionic Liquids (ILs)

Transition metal nanoparticle synthesis in ionic liquids (ILs) can be carried out through thermal reduction[48,49,50,51] or decomposition[52] of metal salts or metal complexes, the photochemical[53,54] or electrochemical-reduction[55,56,57] and recently also by thermal and photochemical decomposition of metal carbonyls.[58,59,60,61,62] In addition, a correlation was found between the molecular volume of the anion in the ionic liquid and the size of the synthesized metal nanoparticles.[48,58,60] (see Chapter 3.1 – 3.6)

The reduction of metal salts by hydrogen gas,[63] photochemical or electroreduction[64] can been used for M-NP synthesis. Iridium nanoparticles were obtained via rapid reduction of [IrCl(cod)]$_2$ (cod = 1,5-cyclooctadiene) in BMIm$^+$PF$_6^-$ with H$_2$ at 75 °C. A similar procedure has also been applied to prepare rhodium, ruthenium[65,66] and nickel[67] nanoparticles.

Silver nanoparticles (Ag-NPs) were prepared through the reduction of silver salts, AgX, by H$_2$ in monophasic ILs and the immediate neutralization of the HX formed through the presence of an imidazole (BMIm) scavenger. (see, Chapter 3.1) To avoid proton (H$^+$/H$_3$O$^+$) incor-poration and impurities in the IL-supramolecular network structure, a cleaner and simple synthesis method was developed. Transition metal carbonyl compounds are attractive starting materials for the synthesis of the most important transition metal nanoparticles (Table 1.2). The synthesis uses easily commercially available metal carbonyl precursors and can readily be expanded to the broad range of other metal carbonyl complexes.

Stable transition M-NPs in ILs can be obtained by simple thermolytic, photolytic decomposition treatment (Chapter 3.2-3.4).[58,59,60] Furthermore, metal carbonyls as educts are available in high purity or can be easily purified, e.g. by sublimation.[68] In contrast to other methods, no further chemicals (like, reduction H_2; CO gas, or $NaBH_4$ and LAH) must be used. Also no scavenger (like, BMIm) is necessary, to avoid proton (H^+/H_3O^+) incorporation in the IL dynamic matrix (see, Chapter 3.1).

Table 1.2: Overview over different mono- and polynuclear transition metal carbonyls precursors.

Cr – Mo – W	Mn – Tc – Re	Fe – Ru – Os	Co – Rh – Ir	Ni – Pd – Pt
$Cr(CO)_6$	$Mn_2(CO)_{10}$	$Fe(CO)_5$ $Fe_2(CO)_9$ $Fe_3(CO)_{12}$	$Co_2(CO)_8$ $Co_4(CO)_{12}$ $Co_6(CO)_{16}$	$Ni(CO)_4$
$Mo(CO)_6$	$Tc_2(CO)_{10}$	$Ru(CO)_5$ $Ru_2(CO)_9$ $Ru_3(CO)_{12}$	$Rh_2(CO)_8$ $Rh_4(CO)_{12}$ $Rh_6(CO)_{16}$	---
$W(CO)_6$	$Re_2(CO)_{10}$	$Os(CO)_5$ $Os_2(CO)_9$ $Os_3(CO)_{12}$	$Ir_4(CO)_{12}$ $Ir_6(CO)_{16}$	---

1.1.4 Gold Nanoparticles (Au-NPs) and Historic Background

Gold is a preferred metal in nanotechnology research for most applications that involve colloidal metals.[69] Nanoparticles of gold have a much longer history than those of any other metal.[70] The *Purple of Cassius* discovered in 1663, formed on reacting stannic acid with chloroauric acid, was a popular purple dye for glass articles and fabrics. It is actually made up of tin oxide and Au nanocrystals.[71] The Romans were first to be adept at impregnating glass with metal particles to achieve different color effects. The *Lycurgus cup*, is the only complete example of a very special type of glass, known as dichroic, which changes colour when held up to the light. (Fig. 1.1.7) The opaque green cup turns to a glowing translucent red when light is shone through it. The glass contains tiny amounts of colloidal gold and silver particles, which give it these unusual optical properties.

Fig.1.1.7: The *Lycurgus cup*, a dichroic glass cup with a mythological scene (Roman, 4th century AD).

Systematic research started with Faraday´s ground-breaking work in 1857 about colloidal gold solutions, followed by the work of *Zsigmondy*[72] (Nobel laureate 1925) and *Turkevich*[73], which describes the "finely divided metallic state" of gold.[74] The importance of gold in the field of nanoscience and nanotechnology stems from its unique stability as a pure metal.[75] Most other metals under ambient conditions are rapidly covered by a passivating oxide film, which make them unsuitable for the fabrication of metallic nanostructures, in which the majority of metal atoms are located at the surface. Most colloidal gold, or gold nanoparticles syntheses have been carried out using the classical Turkevich citrate reduction route[89,75] and its modifications[76,77] in common organic solvents. Thus, typically HAuCl$_4$ is reduced by citric acid which simultaneously functions as a stabilizing agent or reduced by NaBH$_4$ in the presence of thiol or carboxylic acid stabilizers. Yet, for most applications of metal nanoparticles (Au-NPs) it is necessary to transfer them from the synthesis medium into other organic or aqueous solvents or onto surfaces.[79,80] The wealth of size-, shape- and environment-dependent optical properties of gold nanoparticles[78] contributes to the choice of gold in nanotechnological applications in catalysis, chemical and biological sensors, actuators, optical coatings and electronic devices.[79,80]

Still today, the possibility of a controlled and ligand free step by step growth of Au-NPs is not known (Fig. 1.1.8). (see Chapter 3.6)

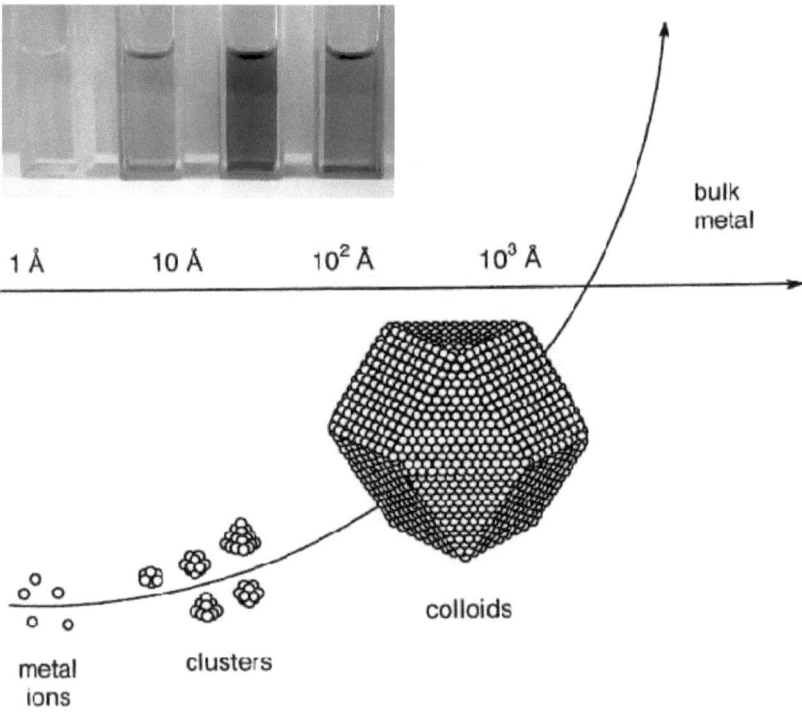

Fig.1.1.8: Schematically growth and typical colour changes during the Au-NPs nucleation and growth.

Chapter 1.2: Industrial Application of Ionic Liquids & Network Properties

1.2.1 Introduction & Applications of Ionic Liquids (ILs) in the Chemical Industry

Room temperature ionic liquids[81] (RTILs) are fluids at room temperature or slightly above – having a melting point below 100°C. ILs are composed entirely of ions, typically large organic cations and small inorganic or organic anions (Scheme 1.2.1). Room temperature ionic liquids (RTILs) have recently attracted considerable worldwide attention as potential alternatives to conventional molecular organic solvents. They have applications in a variety of catalyses, separation technology, electrochemical processes and nanotechnology.

Ionic liquids (ILs) offer the potential for ground-breaking changes to synthesis routes and unit operations in the chemical industry. Essentially salts that are liquid at room temperature, their non-detectable vapor pressures and their unique solvent properties provide the possibility for clean manufacturing. The field of ionic liquids[82] began 1914 with an observation by Paul Wal-den, who reported the physical properties of ethylammonium nitrate ([EtNH$_3$][NO$_3$]; mp 13–14 °C), which was formed by the neutralisation of ethylamine with concentrated nitric acid.[83]

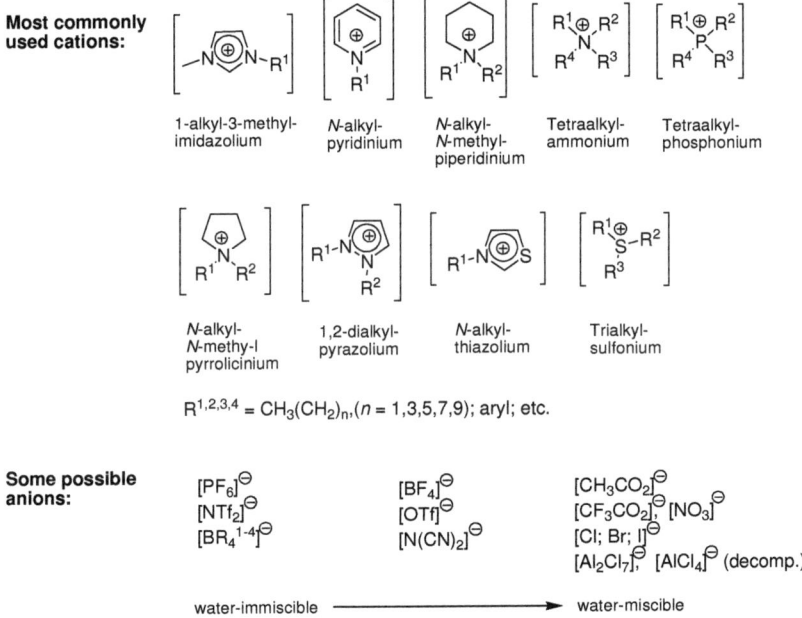

Scheme 1.2.1: Overview over the commonly used ionic liquids (ILs).

These solvents are composed entirely of ions, and strongly resemble ionic melts that may be produced by heating inorganic salts to high temperatures (>> 800°C). They generally consist of a large nitrogen-containing organic cation and a smaller inorganic or organic anion. The asymmetry reduces the lattice energy of the crystalline structure and results in a low melting point salt.[84] There are estimated to be hundreds of thousands of simple ion combinations to make ionic liquids, and a near endless (10^{18}) number of potential ionic liquid mixtures. This implies that it should be possible to design an ionic liquid with the desired properties to suit a particular application by selecting anions, cations, and mixture concentrations. Ionic liquids can be adjusted or tuned to provide a specific melting point, viscosity, density, hydrophobicity, miscibility, etc. for specific chemical systems. Therefore, ionic liquids can legitimately be called "designer" solvents, and offer a freedom and flexibility for process design previously unknown and undreamt.[85]

The components of ionic liquids (ions) are constrained by strong coulombic forces and interactions, and thus exert practically no vapor pressure above the liquid surface. Importantly, the near-zero vapor pressure (non-volatile) property of ionic liquids means they do not emit the potentially hazardous volatile organic compounds (VOC) associated with many industrial solvents during their transportation, handling, and use. (It should be noted, however, that the decomposition products of ionic liquids from excessive temperatures can have measurable vapor pressures.) In addition, they are nonexplosive and non-oxidizing (nonflammable). These characterizations could contribute to the development of new reactions and processes that provide significant environmental, safety, and health benefits compared to existing chemical systems.[86]

The scientific literature reports numerous chemical reactions in which ionic liquids are the media in which the reaction occurs. These include cracking, dissolution, hydrogenation, dimerization, isomerization, oligomerization, and other reactions.[87] The ionic liquids used in a reaction or in catalytic systems are reported to show better activity, selectivity, and stability than traditional systems. They provide better yields, better and more controllable distribution of reaction products, and in some cases faster kinetics. Reactions in ionic liquids also occur at significantly lower temperatures and pressures than conventional reactions, resulting in lower energy costs and capital equipment costs. Ionic liquids can act as both catalyst and solvent.[88] In many systems, the reaction products can be separated by simple liquid-liquid extraction, avoiding energy-intensive and costly distillation.

Many ionic liquids are highly polar and non-coordinating – ideal for catalytic reactions and the synthesis of nanoparticles & nanomaterials. Many are immiscible with water, saturated hydrocarbons, dialkyl ethers, and a number of common inorganic and organometallic precursors – providing flexibility for reaction and separation schemes – and they are nonvolatile even at elevated temperatures, providing a basis for clean manufacturing of nanomaterials and nanoparticles – "green chemistry."[89] By the way, today ionic liquids (ILs) are quite expensive. For a large scale application, the material cost should become more important. The price of a large-scale commercial ionic liquid should be dictated by the price of the cation and anion source. Over a medium term timescale, it would be expected that a range of ionic liquids becomes commercially available for 25–50 € per liter on a ton scale.[125]

1.2.2 Ionic Liquids Involved in Important Large Scale Industrial Processes

All of the industrial giants, including BASF, MERCK and EVONIK have done the most effort to implement ionic liquid technology. They possess the largest patent portfolio and have the broadest range of applications.

1.2.2.1 The BASF BASIL™ process.
Arguably, the most successful example of a large scale industrial process using ionic liquid technology is the BASIL™ (**BASIL** = **B**iphasic **A**cid **S**cavenging utilizing **I**onic **L**iquids) process.[90] This first commercial publicly announced process was introduced to the BASF site in Ludwigshafen, Germany, in 2002.

In 2002 BASF established the first dedicated industrial-scale ionic liquid based process. The BASIL™ process is used for the synthesis of alkoxyphenylphosphines which are important raw materials in the production of BASF's Lucirines® (Scheme 1.2.2), products that are used as photoinitiators to cure coatings and printing inks by exposure to UV light.

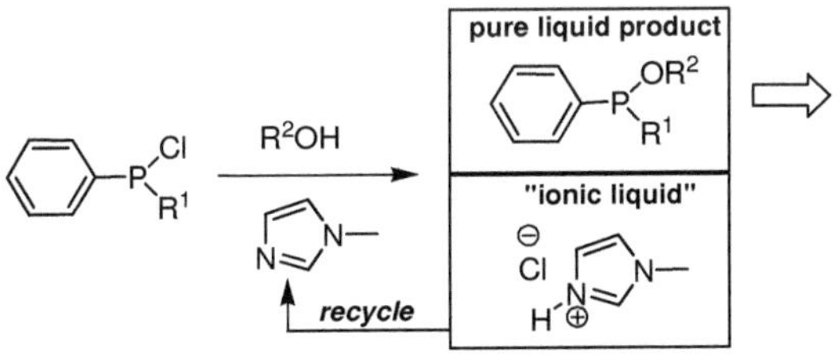

Scheme 1.2.2: The use of the BASF BASIL™ process in the production of Lucirines®.[91]

The ionic liquid is acting as an auxiliary in the BASIL™ process, benefits are:

- no handling of solids
- better heat transfer
- higher chemical yield
- higher space-time-yield
- lower investment cost
- higher sustainability of the process

In the original process, triethylamine was used to scavenge the acid that was formed in the course of the reaction, but this made the reaction mixture difficult to handle as the waste by-product, triethylammonium chloride formed a dense insoluble paste.

Replacing triethylamine with 1-methylimidazole, results in the formation of 1-methyl-imidazolium chloride (HMIm$^+$Cl$^-$), an ionic liquid which has a melting point of about 75°C and separates the reaction mixture as a discrete phase. After the reaction two clear liquid phases occur that can easily be separated. The upper phase is the pure product – no solvent is needed anymore - the lower phase the pure ionic liquid (Fig. 1.2.2 and Scheme 1.2.2). HMIm$^+$Cl$^-$ can be switched on and off just by protonation and deprotonation (Scheme 1.2.2). This is crucial when recycling and purification of the ionic liquids is considered.

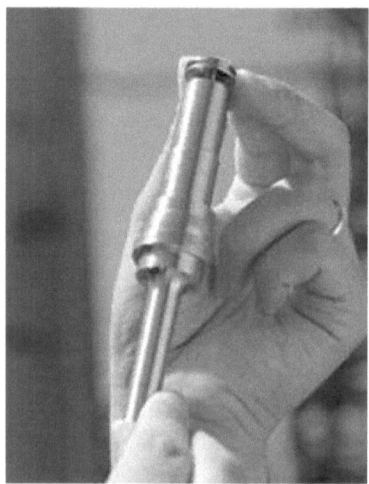 

Figure 1.2.1: The BASIL™ *jet stream reactor*.[92]   Figure 1.2.2: The BASIL™ two phase reaction.[93]

Further investigations revealed that BASIL™ is not restricted to phosphorylation chemistry but is a general solution to all kinds of *acid scavenging*. Acylations and silylations have been run successfully as well as an elimination reaction. BASIL™ is also applicable to extractive acid removal from organic phases for example for the purpose of purification.

Today the BASIL™ process is run in a little *"jet stream reactor"* (see **Fig. 1.2.1**) compared to the initial process (Fig. 1.2.2); the space-time yield is increased from 8 kg m^{-3} h^{-1} to 690,000 kg m^{-3} h^{-1}, and the yield increased from 50% to 98%. 1-Methylimidazole is recycled, via base decomposition of 1-H-3-methylimidazolium chloride (**HMIm$^+$Cl$^-$**), in a proprietary process.[94] The reaction is now carried out at a *multi-ton scale*, proving that handling large quantities of ionic liquids is practical.[95]

1.2.2.2 Dimersol[97] and Difasol Process[96]:

The Dimersol process, based on traditional technology, consists of the dimerisation of alkenes, typically propene (Dimersol-G) and butenes (Dimersol-X) to the more valuable branched hexenes and octenes.[96] This is an important industrial process, with thirty-five plants in operation worldwide, each plant producing between 20,000 and 90,000 tonnes per year of dimer, with a total annual production of 3,500,000 tonnes.[97] The longer-chain olefins produced in the dimerisation process are usually hydroformylated to alcohols (e.g. isononanols): isononanols are then converted into dialkyl phthalates, which are used as poly(vinylchloride) plasticisers.[98](see, Scheme 1.2.3)

The dimerisation reaction is catalysed by a cationic nickel complex of the general form [Ni(PR$_3$)(CH$_2$R)][AlCl$_4$] and is commonly operated without solvent. The use of chloro-aluminate(III) ionic liquids as solvents for these nickel-catalysed dimerisation reactions has been developed and pioneered at IFP (France), especially by Nobel laureate Yves Chauvin and Héléne Olivier Bourbigou.[99, 100] The reaction can be performed as a biphasic system bet-ween -15 °C and 5 °C, as the products form a second layer that can be easily separated and the catalysts remains selectively dissolved in the ionic liquid phase. The activity of the catalyst is much higher than in both solvent-free and conventional solvent systems, and the selectivity for desirable dimers is enhanced. This process has been patented as the Difasol Process.[96]

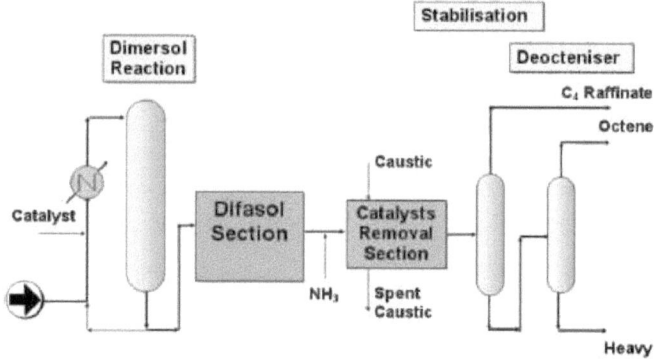

Scheme 1.2.3: Dimersol and Difasol process.[108]

1.2.2.3 BASF Cellionic™: Solutions of Cellulose

One of the most remarkable properties of ionic liquids is their ability to dissolve cellulose. Cellulose is the earth's most widespread natural organic chemical and, thus, highly important biopolymer as a bio-renewable resource at a total volume of about 700 billion tons. But even out of the 40 billion tons nature renews every year, only 0.1 billion tons are used as feedstock for further processing. For example, making cellulose fiber from so-called dissolving pulp currently involves the use, and subsequent disposal, of great volumes of various chemical additives. A total of some 600,000 metric tons of carbon disulfide (CS_2) is consumed each year for this application. For each ton of cellulose fiber, there are more than two tons of waste substances. Major volumes of waste water are also produced for process reasons and need to be disposed of. The Cellionic™ opens up a broad range of opportunities which so far could not be investigated, as solutions of cellulose in chemically inert solvents simply did not exist.[101]

These processes can be greatly simplified by the use of ionic liquids, which serve as solvents and are nearly entirely recycled. This can clearly reduce the amount of auxiliaries needed. Cellulose in ionic liquids gives *real physical solutions*. The crystallinity of the cellulose raw material fully disappears on dissolving it in the ionic liquid. Ethyl-Methyl-Imidazolium Acetat ($EMIm^+OAc^-$) can dissolve up to 25 wt.% of cellulose. The cellulose solutions show excellent long-term stability even at elevated temperatures without significant decrease of the average degree of polymerization (DP). They do not contain gel particles and are air-stable. BASF has now launched a series of solutions of cellulose in ionic liquids under the brand name Cellionics™.[102]

1.2.2.4 Electro deposition of Metals and Semiconductors:

Various metals and their alloys like, aluminium, indium, antimony, copper, silver, tellurium, cadmium, palladium, gold, nickel and cobalt can be electrodeposited to nanometer thin films from various inorganic salt sources in different ionic liquids.[103] Moreover, some ionic liquids, like $BMIm^+BF_4^-$, $BMIm^+OTf^-$ or $BMIm^+NTf_2^-$ have an extremely large electrochemical window of more than 5 V, and hence they give access to the electrodeposition of many metals and semiconductors, such as Ta, Ti, Si, Al and Ge.[104]

Firstly, aluminium and other metals were electrodeposited was from chloroaluminate ionic liquids on different substrates by several authors using classical electrochemical methods, such as cyclic voltammetry, potential step experiments and ex situ techniques.[105] In all cases, metal deposition was observed in an acidic regime and the quality of the deposits was reported to be superior to those obtained from organic solutions. The deposition on substrates like glassy carbon, tungsten and platinum is preceded by a nucleation step and the deposition is electrochemically quasireversible. The thickness of the films and pattern can be easily regulated through the used electro deposition current.[106]

The quality of electrodeposited Ge films can be seen in Fig. 1.2.3 by SEM. An applicative example of the electrodeposition of aluminium on an iron nail in ionic liquids can be seen in Fig. 1.2.4.

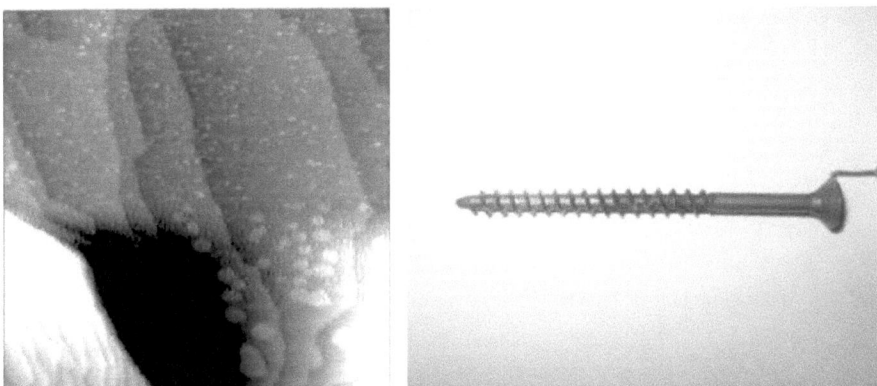

Fig. 1.2.3: Electrodeposited Ge films, SEM picture.[106] Fig. 1.2.4: Electrodeposition of Al on an iron nail.[107]

1.2.2.4 Applications and future directions of ionic liquids (ILs):

The field of ionic liquids is growing at a rate that was unpredictable even five years ago. The range of commercial applications is quite staggering; not just in the number, but in their wide diversity, arising from close cooperation between academia and industry.[108] Even though, concepts and paradigms of ionic liquids are quite new, and still not really accepted in the wider community and chemical industry: it is hard for a conservative scientist to throw away the concepts of molecular solvents and syntheses media, and if chemists are conservative, then chemical engineers are even more so.[87]

Scheme 1.2.4 gives a close examination of the potential application of ionic liquids, thus, this reveals that these applications are not haruspicy, but a natural extrapolation of where we are now. It is thus not really a prediction but an expectation.[81,108]

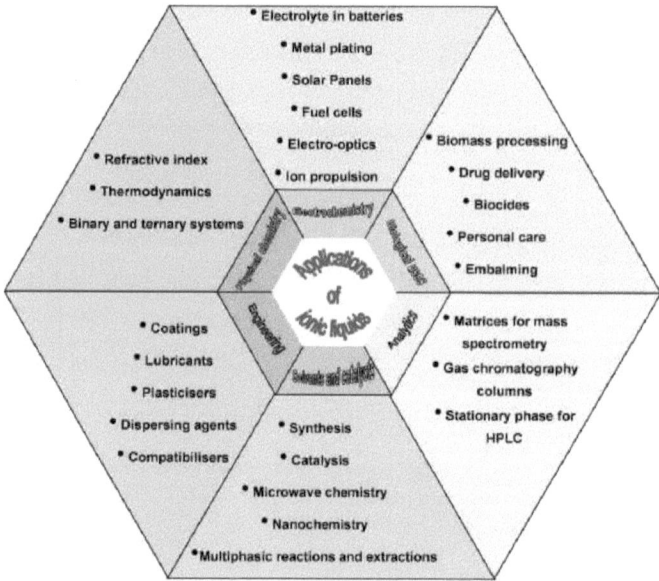

Scheme 1.2.4: Applications of ionic liquids.[81,108]

Further applications of ILs will be found[109,110] and even more petrochemical and industrial large scale applications will be realised. However, the most exciting application will be based on their predictable properties for the synthesis of defined nanoparticles & materials. As ionic liquids can, in principle, replace conventional liquids and syntheses media wherever they are used today, we have barely scratched the surface of the possibilities. Industrial large scale applications of ILs in the next few years will maybe open the common use of these "salts".

1.2.3 Network Properties of Ionic Liquids (ILs)

Ionic liquids are now very popular and enjoy a plethora of applications in various domains of physical sciences. For example, they are used as "solvents" for organic, organometallic syn-theses and catalyses[111], as electrolytes in electrochemistry, in fuel[112] and solar cells[113], as lub-ricants[114], as a stationary phase for chromatography[115], as matrices for mass spectrometry[116], supports for the immobilization of enzymes[117], in separation technologies[118], as liquid crystals[119] templates for the synthesis of mesoporous[120] and nano-materials[121] and ordered films[122], materials for embalming and tissue preservation[123], etc. Not surprisingly an impressive number of specialized reviews[124] and books[125] have recently appeared dealing with their syntheses, physical-chemical properties and applications.

The physical-chemical properties can be attributed to their supramolecular network character. An overview in the X-ray studies reported in the last years on the structure of 1,3-dialkyl-imidazolium (DAIm) salts[126] reveals a typical trend: *"they form in the solid state an extended network of cations and anions connected together by hydrogen bonds."* (Scheme 1.2.5)[127]

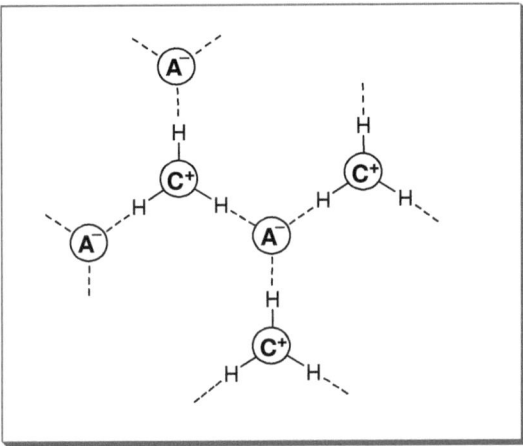

Scheme 1.2.5: Two-dimensional simplified solid-state model of the polymeric supramolecular structure of 1,3-dialkyl imidazolium ionic liquids showing the hydrogen bonds between the imidazolium cation (C) and the anions (A) (one cation is surrounded by three anions and vice-versa). Note that even in the case of $B(Ar)_4$ anions the network is formed through relatively strong C-H···π hydrogen bonds.[128]

The packing arrangement is always constituted of one imidazolium cation surrounded by at least three anions and in turn each anion is surrounded by at least three imidazolium cations (Scheme 1.2.5). The strongest hydrogen bond always involves the most acidic H^2 of the Imidazolium cation (pKa=23.0 for the 1,3-dimethyl imidazolium cation)[129] followed by the other two hydrogens (H^4 and H^5) of the imidazolium nucleus and/or the hydrogens of the N-α-carbon atom chains (H^6, H^7 and H^8). These bonds possess properties of weak to moderate hydrogen bonds – they are mostly electrostatic in nature - (H⋯X bond lengths > 2.2Å; C-H⋯X bond angles between 100°-180°). (Scheme 1.2.6)[130]

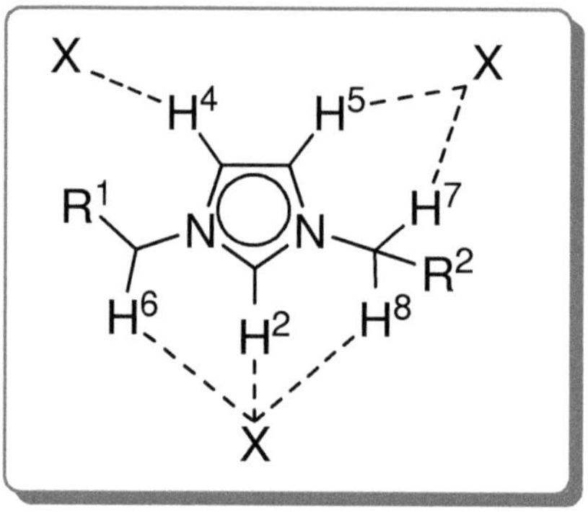

Scheme 1.2.6: Network structure detail from the network structure of Scheme 2.5 for imidazolium H bonded to X anions (H⋯X bond lengths > 2.2Å; C-H⋯X bond angles between 100°-180°).

Although the number of anions that surround the cation (and vice-versa) can change depending upon the anion size and type of the N-alkyl imidazolium substituents, the structural trend of one imidazolium cation hydrogen-bonded to at least three anions and one anion hydrogen-bonded to at least three cations is a general trend in imidazolium salts.

The three dimensional arrangements of the imidazolium ionic liquids in the solid phase are generally formed trough chains of the imidazolium rings (Fig. 1.2.5). In some cases there are typical π-π stacking interactions among the imidazolium rings and in the case of 1-alkyl-3-methylimidazolium salts relatively weak C-H⋯π interactions via the methyl group and the imidazolium ring-π system can be also found. This molecular arrangement can generate channels in which the spherical anions are accommodated as chains (Figure 1.2.5). This structural pattern depends on the anion geometry, and the internal arrangements along the imidazolium columns vary with the type of the N-alkyl substituents. This structural pattern is a general trend for the solid and liquid phase.[131]

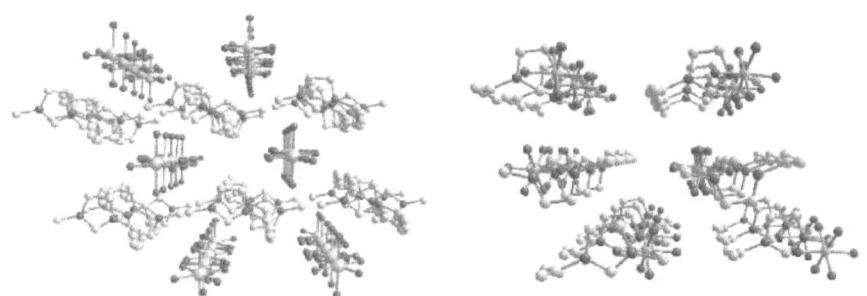

Figure 1.2.5: Packing diagram for a single crystal X-ray structure of 1-*sec*-butyl-3-methy-limidazolium cation as SbF_6^- (left) and PF_6^- salt (right).[88]

"Pure" 1,3-dialkylimidazolium ionic could be described as *well-organized hydrogen-bonded polymeric supramolecules* of the type $\{[(DAIm)_x(X)_{x-n}]^{n+} [(DAIm)_{x-n}(X)_x]^{n-}\}_n$ where DAIm is the 1,3-dialkylimidazolium cation and X the anion.[127]

Moreover there is now much evidence indicating that the supramolecular organization, recog-nition and interaction can not be maintained when they are mixed with other substances.[132] Therefore, this incorporation of other molecules and nanoparticles in the ionic liquid network causes changes to the physicochemical properties of these materials and can cause the for-mation of polar and nonpolar nanoregions. These inclusion compounds can involve molecules (such as arenes)[133], ions (such as charged transition-metal complexes)[134], macromolecules (such as enzymes[135] or δ cellulose[136]) or nanoparticles (such as transition-metal nano clusters)[137], and the stabilisation of this process is mainly due to the electronic and steric effects provided by the nanostructures of *"imidazolium cation/anion clusters"* in the ionic liquid network structure. (Scheme 1.2.7 and 1.2.8)

Transition-metal nanoparticles can be generated by simple reduction of metal salts or by controlled decomposition of organometallic compounds "dissolved" in imidazolium ionic liquids ($DAIm^+X^-$).[138] Using this *general network approach* various stable nanometric transition-metal particles with defined size and size narrow distribution can be prepared.

The interaction between ionic liquids and metal nanoparticles creates a protective layer com-posed of imidazolium aggregate anions located immediately adjacent to the nanoparticle sur-face – providing the Coulombic repulsion – and counter cations that provide the charge ba-lance, which is quite close to a DLVO (Derjaugin-Landau-Verwey-Overbeek) type stabili-sation[139]. However, the pure DLVO model can not completely explain the stabilisation proper-ties of imidazolium ionic liquids towards various metal nanoparticles since it treats counter-ions as mono-ionic point charges and was not designed to account for sterically stabilised systems[140] Moreover, some of the nanoparticles stabilisation in ionic liquids may be provided by the presence of surface attached carbenes like in the case of Ir(0) nanoparticles.[141]

The formation and stabilisation of nanoparticles in these fluids occurs with the reorganisation of the hydrogen bond network (Scheme 1.2.5 and 1.2.6) and generates nanostructures with polar and nonpolar nano-regions (Scheme 1.2.7 and 1.2.8).

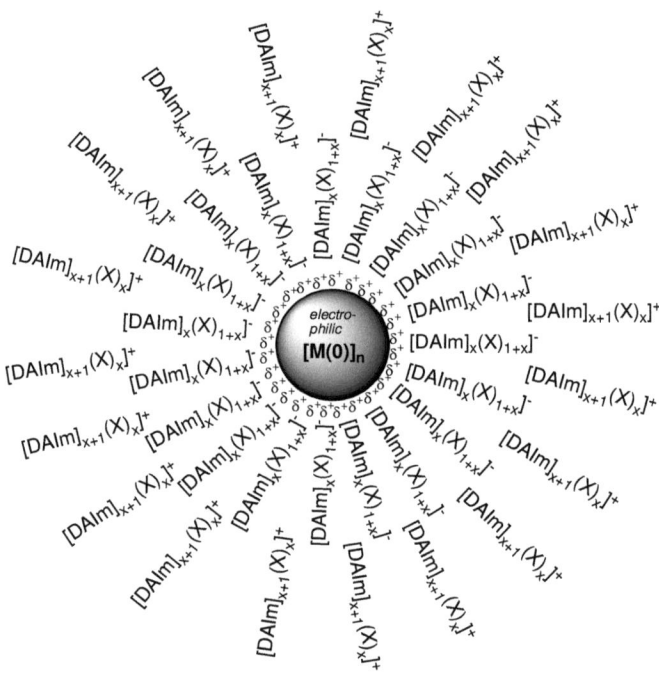

Scheme 1.2.7: The steric type stabilisation build up by the anionic and cationic supramolecular aggregates of the type $[(DAIm)_x(X)_{x-n}]^{n+} [(DAIm)_{x-n}(X)_x]^{n-}$ in the ionic liquid network.[143]

The ionic liquid forms a protective layer surrounding the transition-metal nanoparticles in ways that surface, depend on the type of the anion, suggesting the presence of semi-organised anionic species composed of supramolecular aggregates of the type $[(DAIm)_{x-n}(X)_x]^{n-}$. This structural organisation is similar to that already observed in solid and liquid phases and in solution of imidazolium salts. This protective layer is probably composed of imidazolium aggregate anions located immediately adjacent to the nanoparticle surface, quite close to "DLVO-type" stabilisation (see Scheme 1.2.8).[127]

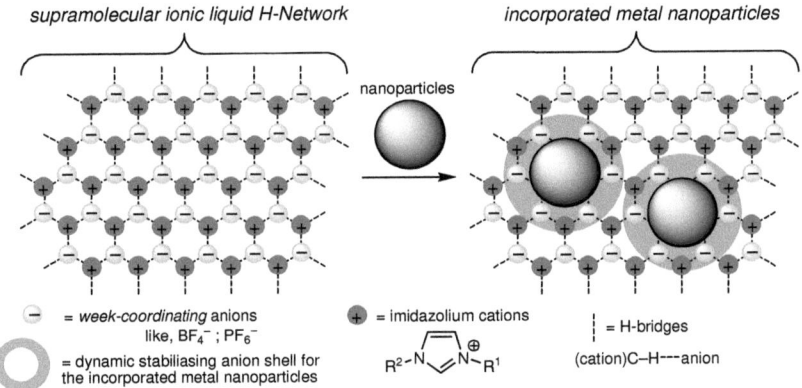

Scheme 1.2.8: Network structure and incorporation of metal nanoparticles in the IL network.[189]

The anions which are located immediately adjacent to the nanoparticle surface were investigated by XPS analysis on Pd nanoparticles in $BMIm^+PF_6^-$.

The XPS analysis showed the presence of palladium, fluorine, carbon and a small contribution of phosphorus. It is clear that the F and P signals indicate that the Pd nanoparticles contain residues from the ionic liquid. No other impurities were detected within the sensitivity of the technique. Figure 1.2.6 shows the XPS signal of the Pd 3d and F 1s regions. The Pd 3d spectrum indicates the presence of two chemical states of Pd at the nanoparticle surface with distinct binding energies; the main contribution is related to Pd(0) (Pd–Pd bonds, Pd 5/2 at 335 eV see, Fig. 1.2.6) and the other corresponding to Pd–F interactions (Pd 5/2 at 336.7 eV; see, Fig. 1.2.6). It was found that the relation between the Pd–Pd and the Pd–F areas in the spectrum is about 2.8, which corresponds to a 74% contribution of Pd–Pd bonds and 26% of Pd–F interaction on the particle surface. This result is a strong indication that the effective interaction of the ionic liquid anion with the metal surface is responsible for the stabilisation of the nanoparticles.[142]

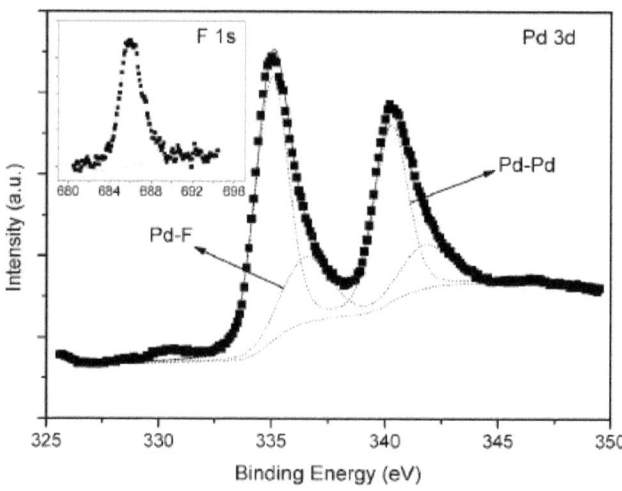

Fig. 1.2.6: X-ray photoelectron spectrum of Pd^0 NPs synthesised from $Pd(acac)_2$ in $BMIm^+PF_6^-$.[142]

Furthermore ^{19}F-NMR shows dynamic anion stabilisation behaviour in the weak coordinating IL-network system. (see, Chapter 3.5.4) A concentration dependent shift was observed in our experiments, which is quite small for ^{19}F-NMR chemical shifts (0.02 ppm), but these findings support the idea that there are interactions between the fluorine atoms of the BF_4^- anions and the surface of the gold nanoparticles. Fast exchange will occur between BF_4^- in the vicinity of Au nanoclusters and the bulk ionic liquid, so on the NMR timescale there are no different resonances but a small effect on the chemical shift of the average signal. These observations contribute to the model which predicts that anions are the primary source of stabilization for electrophilic metal nanoclusters. This supports the hypothesis that weakly-coordinating fluorous anions can contribute to the stability of transition-metal nanoclusters in ILs.[215]

Therefore, together with the *electrostatic stabilisation* provided by the intrinsic strong columbic forces and interaction of the ionic liquid, a *steric type stabilisation* can also be envisioned in the ionic liquid network due to the presence of anionic and cationic supramolecular aggregates of the type $[(DAIm)_x(X)_{x-n}]^{n+}$ $[(DAIm)_{x-n}(X)_x]^{n-}$ where DAIm is the 1-*n*-butyl-3-methyl-imidazolium cation ($BMIm^+$) and X is the anion. These supra-molecular aggregates may be even present in the cases in which the stabilisation may also involve surface attached carbenes like in the case of Ir(0) nanoparticles as transient species.[143]

The sole network model in Scheme 1.2.8 allows for a lot of tuning mechanisms. A difference in size and shape of different cation and anion molecular volumes, (Tab. 1.2.1) allows a broad variation for the creation of new different imidazolium hydrogen-bonded network types.

Table 1.2.1: Different Cation/Anions in ILs and their molecular volume.

Cation$^+$ / molecular Volume $V_{IL\text{-}cation}$ (nm^3)	Anion$^-$ / $V_{IL\text{-}anion}$ (nm^3)
[EMIm]$^+$ (Ethyl-Methyl-Imidazolium) / 0.156 ± 0.018	[BF$_4$]$^-$ / 0.073 ± 0.009
[C$_3$MIm]$^+$ (Propyl-Methyl-Imidazolium) / 0.178 ± 0.028	[PF$_6$]$^-$ / 0.109 ± 0.008
[BMIm]$^+$ (Butyl-Methyl-Imidazolium) / 0.196 ± 0.021	[OTf]$^-$ / 0.131 ± 0.015
[BtMA]$^+$ (Butyl-tri-Methyl-Ammonium) / 0.198 ± 0.013	[NTf$_2$]$^-$ / 0.232 ± 0.015

The modelling of different network types allows a definite environment for the nanoparticle nucleation and growth. In our studies we could reproducibly show that the network structure, and especially the anion composition in the network, posses an important influence on nano-particle size generation.[48,58] Furthermore, gas solubility and the solubility of different solvents can be described through these network systems. Studies of organic solute behaviour in ILs are based on liquid/liquid separations which may facilitate a molecular level understanding of the partitioning mechanisms for neutral and ionic solutes, providing a predictive tool for their behaviour in ILs.[125]

However, the wide range of possible anions and cations with different substituents create the possibility of tailoring ILs for the synthesis of different transition metal nanoparticles, and the use of such dispersion for reactions like hydrogenation, which involves gases, is an exciting option.[125] Therefore, the ionic liquid acts as a weak coordinating *"supramolecular template"* in the generation of nanoparticles (Scheme 1.2.9). Depending, on the molecular anion volume of the ionic liquid (Table 1.2.1) we could generate spherical nanoparticles of different sizes. (see, Chapters 3.1-3.5).[48,58,60]

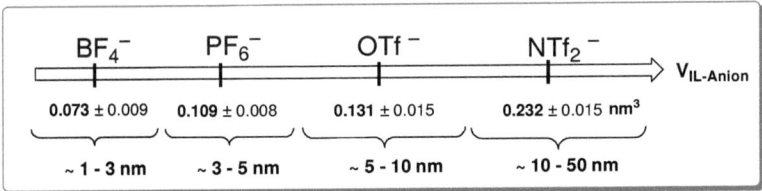

Scheme 1.2.9: Size prediction for synthesized metal-containing nanoparticles in dependence on the ionic liquid anion.[48]

Chapter 2: Assignment

The goal of this Ph.D. project was to combine research in transition Metal Nanoparticles (M-NPs) with Ionic Liquids (ILs). The intrinsic properties of ILs, such as their ionic charge, high polarity, high dielectric constant, high thermal stability and low reactivity were seen as advantageous for use as a nanosynthetic medium, or template, to prepare transition M-NPs.

To find a reproducible, widely applicable, and facile method for synthesising different transition M-NPs (e.g. Cr, Mo, W, Fe, Ru, Os, Co, Rh, Ir, Ag and Au) in ILs was the main aim for our research. Our interest was to find new, new efficient routes for the synthesis of transition M-NPs in ILs, which were expected to function as a non-surfactant species, or as a weakly coordinating supramolecular network, for the kinetic stabilization of the M-NPs, so that no extra stabilizing or capping ligands were needed.

Different parameters in the synthesis procedure, such as the metal source (M-NP precursor), type of IL, and temperature, would be varied so as to control the size of the M-NPs. The size and characteristics of the resulting M-NPs would be analyzed by standard methods, like transmission electron microscopy, X-ray powder diffraction or dynamic light scattering.

The resulting "naked" M-NPs, with or without the IL, should be tested in model reactions, for example hydrogenations, C-C couplings or oxygenations, for their catalytic application and possible repetitive utilization. A capacity for broad surface functionalization should be shown, e.g. by the introduction of organic capping molecules like thiolglycolic acid and *n*-decanethiol, and transfer of M-NPs from the IL to polar and non-polar organic solvents. In addition the deposition onto surfaces from ionic liquids (ILs), such as carbon nanotubes, graphite, or different polymers should be tested.

Furthermore, spectroscopic and theoretical Density Function Theory (DFT) investigations would be initiated with different corporation partners to understand the interaction and stabilization mechanisms of dispersed transition M-NPs in ILs. It should be interesting to know which component of the ionic liquid, anion or cation, has the greatest effect on the stability of the dispersed electrophilic transition metal nanoclusters M-NPs. A clear understanding of possible interactions and stabilization mechanisms will bring new and interesting insights into the field of ILs and transition M-NPs.

Chapter 3: Results and Discussion

Introduction (Synthesis and Stabilization of Transition Metal Nanoparticles in ILs)

Transition Metal nanoparticles (M-NPs) posses a huge importance for technological applications in several areas of science and industry, including catalysis or chemical sensors.[144,145] Syntheses and applications of transition-metal nanoparticles are of contemporary interest in several areas of science.[146]

For example, silver nanoparticles (Ag-NPs) supported on Al_2O_3[147] or titanium silicate zeolite[148] are of industrial importance for the gas phase oxidation of olefins with molecular oxygen from air to their epoxides. In particular, Ru, Ir and Rh nanoparticles are used for olefin hydrogenation reactions.[149] Furthermore, W and Mo nanoparticles can be used for olefin metathesis reactions.[175] The controlled and reproducible synthesis of defined and stable metal nanoparticles (M-NPs) is of very high importance for different fields of applications.[150,151,152,153,154,155]

An intrinsic problem of M-NP synthesis from MX_n salts with H_2 is the formation of strongly acidic HX (X e.g. Cl, NO_3, BF_4, PF_6, O_3SCF_3). An acidic reaction medium destabilizes M-NPs and leads to their clustering (Chapter 3.1).[156] N^1,N^1,N^8,N^8-teramethylnaphthalene-1,8-diamine was present as a proton sponge in the preparation of Ir nanoparticles via reduction of $[Ir(cod)(CH_3CN)_2]BF_4$ under H_2.[157] To avoid proton (H^+/H_3O^+) incorporation and impurities in the IL-supramolecular network structure, a cleaner and simple synthesis method was developed. Through the use of transition metal carbonyl compounds (see, Table 1.2), which are attractive as starting materials most of the important transition metal nanoparticles could be synthesized. Stable transition M-NPs in ILs can be obtained by simple and clean thermolytic and photolytic decomposition treatment (Chapter 3.2-3.4).[58,59,60]

However, transition-metal nanoparticles (M-NPs) in ionic liquids (ILs) are only kinetically stable because the formation of bulk metal is thermodynamically favored. Yet, in the absence of strongly coordinating protective ligand layers, M-NPs in ILs should be effective catalysts. The IL network contains only weakly coordinating cations and anions as stabilizers that bind less strongly to the metal surface than other anions or capping ligands.

Ionic liquids (ILs) as a "*nanosynthetic template*" stabilize metal nanoparticles on the basis of their high ionic charge, high polarity, high dielectric constant and supramolecular network.[158] According to DLVO (Derjaugin-Landau-Verwey-Overbeek) theory[159] ILs provide an electro-static protection in the form of a "*protective shell*" for M-NPs, so that no extra stabilizing molecules or organic solvents are needed.[160]

Ionic liquids stabilize metal nanoparticles on the basis of their high ionic charge, high polarity, high dielectric constant and supramolecular network.[161,162,163,164] According to DLVO (Derjaugin-Landau-Verwey-Overbeek) theory,[165,166] ILs should provide an *electrosteric* protection in the form of a "*protective shell*" for M-NPs, so that no extra stabilizing molecules or organic solvents are needed. Here we note that DLVO theory treats anions as ideal point charges. Real-life anions with a molecular volume would be classified as "*electrosteric* stabilizers". However, the term "*electrosteric*" is contentious and ill-defined.[167] The stabilization of metal nanoclusters in ionic liquids could, thus, be attributed to "extra-DLVO" forces [167] including effects from the network properties of ionic liquids like hydrogen bonding, hydrophobic effects and steric interactions. However, the DLVO theory predicts that the anionic charges should be the primary source of stabilization for the electrophilic metal nanocluster. Nanoparticles must be stabilized in order to prevent their agglomeration or aggregation which eventually leads to the formation of small metal particles. M-NPs (*core*) are considered stabilized in ILs by the formation of "*protective*" anionic and cationic layers (*shells*) around them in a "*core-shell system*".[168,169,170,171] According to DLVO theory the first inner shell must be anionic:[165] Then the IL anion will have the highest influence on the size and electrostatic stabilization of the electrophilic transition metal nanoparticles.

The ionic liquid represents an excellent medium for the formation of electrophilic transition metal nanoclusters with, in most cases, a small size and size distribution under mild conditions.

Chapter 3.1: The First Correlation of Nanoparticle Size Dependent Formation with the Ionic Liquid Anion Molecular Volume

Silver nanoparticles (Ag-NPs) were prepared through the reduction of silver salts, AgX, by H_2 in monophasic ILs and the immediate neutralization of the HX formed through the presence of an imidazole scavenger. The imidazole and anion X is chosen, such as, to lead to the formation of an ionic liquid, preferably identical to the one already employed as the reaction medium. Even if the ionic liquid formed is only similar it will be miscible with the main IL. In BASF's BASIL^{TM} process (*B*iphasic *A*cid *S*cavenging utilizing *I*onic *L*iquids), see Chapter 1.2, 1-alkylimidazoles are used to scavenge acid byproducts from organic chemical processes in order to prevent decomposition of the primary reaction product or to prevent unwanted side-reactions.[172] Furthermore, we show that the average silver M-NP particle size depends on and increases with the molar volume of the IL anion.

The silver precursors $AgBF_4$, $AgPF_6$ or AgOTf were dissolved, Ag_2O was suspended under argon in dried and deoxygenated ionic liquids (Scheme 3.1.1) and reacted with H_2 (4 atm, 85 °C) The reduction was carried out in the absence and presence of *n*-butyl-imidazole scavenger (Bim) in a stainless steel reactor (Scheme 3.1.2).

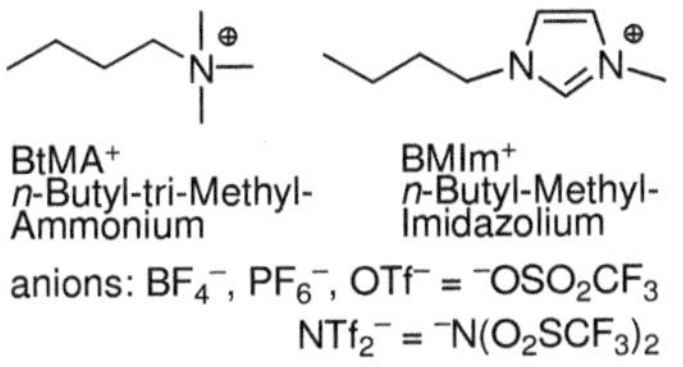

BtMA⁺
n-Butyl-tri-Methyl-Ammonium

BMIm⁺
n-Butyl-Methyl-Imidazolium

anions: BF_4^-, PF_6^-, $OTf^- = {}^-OSO_2CF_3$
$NTf_2^- = {}^-N(O_2SCF_3)_2$

Scheme 3.1.1: Used ionic liquids (ILs).

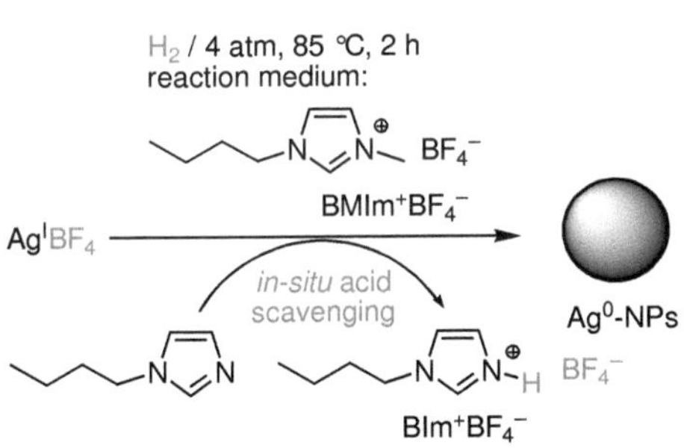

Scheme 3.1.2: Formation and stabilization of silver nanoparticles (Ag-NPs) by hydrogen reduction of $AgBF_4$ with the imidazole scavenging process in BMIm+BF_4^-. The ionic liquid formed from the scavenging process should be similar to the main IL solvent.

In the absence of the BIm scavenger the Ag-NP particle size distribution is very broad, with a range of several 10 nm or even 100 nm (Table 3.1.1, Fig. 3.1.1). This can be reasoned by proton (H^+/H_3O^+) incorporation in the IL dynamic matrix (see below).[48] Also, the Ag-NP dispersion prepared without a scavenger is unstable as evidenced by clearly visible metal particle precipitation within 1-2 hours after reduction. In the presence of the BIm scavenger and soluble silver salts the distribution of the Ag nanoparticles lies largely within 10 nm (Table 3.1.1, Fig. 3.1.2) and the dispersion is stable up to 3 days under argon.

Table 3.1.1: AgNP size and distribution from different ionic liquids, silver precursors and with or without scavenger.

	Ionic liquid [a]	$V_{IL\text{-}Anion}$ / nm^3 [185]	Silver precursor	Ag-NP median (min-max) diameter / nm, standard deviation σ [b]
	without scavenger – soluble silver salts			
1	BMIm$^+$BF$_4^-$	0.073 ± 0.009	AgBF$_4$	66 (0.9-267) [c]
2	BMIm$^+$PF$_6^-$	0.109 ± 0.008	AgPF$_6$	9 (0.8-215) [c]
3	BMIm$^+$OTf$^-$	0.131 ± 0.015	AgOTf	7.7 (0.6-27) [c]
	without scavenger – insoluble silver precursor			
4	BMIm$^+$BF$_4^-$	0.073 ± 0.009	Ag$_2$O	5.7 (1.6-42) [c]
5	BtMA$^+$Tf$_2$N$^-$	0.232 ± 0.015	Ag$_2$O	8.6 (0.8-41) [c]
	with BIm scavenger – soluble silver salts			
6	BMIm$^+$BF$_4^-$	0.073 ± 0.009	AgBF$_4$	2.8 (1.2-4.7), σ = 0.8
7	BMIm$^+$PF$_6^-$	0.109 ± 0.008	AgPF$_6$	4.4 (2.0-9.8), σ = 1.3
8	BMIm$^+$OTf$^-$	0.131 ± 0.015	AgOTf	8.7 (3.5-18.4), σ = 3.4
9	BtMA$^+$NTf$_2^-$	0.232 ± 0.015	AgOTf [d]	26.1 (12.0-40.6), σ = 6.4
	with BIm scavenger – insoluble silver precursor			
10	BMIm$^+$BF$_4^-$	0.073 ±0.009	Ag$_2$O	6.5 (1.6-37) [c]

[a] BMIm$^+$ = n-butyl-methyl-imidazolium, BtMA$^+$ = n-butyl-trimethyl-ammonium, BIm = n-butyl-imidazole; [b] from TEM measurements, statistical evaluation of the total sample pictures; [c] very broad distribution curve, slowly tailing off to large particles, thus, no standard deviation given; [d] scavenging with 1-methyl-imidazole.

Under the reaction conditions the median silver nanoparticle size ranges from 3 to 26 nm with uniform size distribution. The scavenging process is crucial for the formation of well defined, stable and finely dispersed Ag nanoparticles in the IL matrix. The starting material silver(I) oxide, Ag$_2$O is insoluble in ILs and upon reduction gives Ag nanoparticles with a broad size distribution (Table 3.1.1, Fig. 3.1.2). Hence, dissolution of the metal starting material in ILs is important for the synthesis of Ag-NPs with narrow size distribution in ionic liquids.

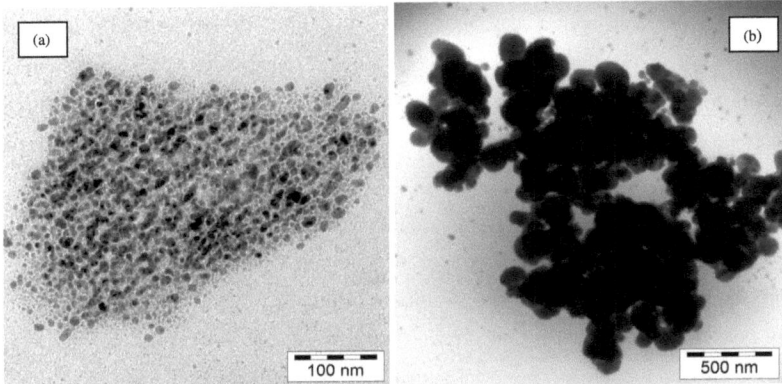

Figure 3.1.1: Ag nanoparticles from (a) AgOTf in BMIm$^+$OTf$^-$ and (b) AgBF$_4$ in BMIm$^+$BF$_4^-$ produced without scavenger (entry 1 and 3 in Table 3.1) (TEM photographs).

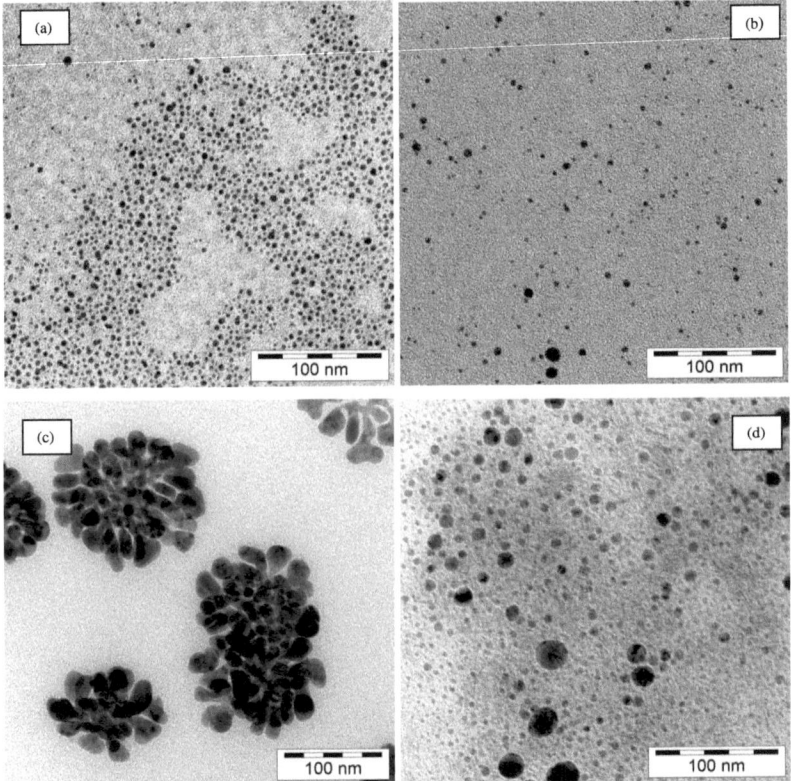

Figure 3.1.2: Ag nanoparticles from (a) AgBF$_4$ in BMIm$^+$BF$_4^-$, (b) AgPF$_6$ in BMIm$^+$PF$_6^-$ and (c) AgOTf in BtMA$^+$NTf$_2^-$ produced with BIm and MIm as scavenger (entry 6,7 and 9 in Table 3.1); (d) Ag nanoparticles from Ag$_2$O in BMIm$^+$BF$_4^-$ without an scavenger (TEM photograph).

Ag nanoparticles prepared with a scavenger in this medium are finely dispersed in the IL matrix (Fig. 3.1.2). Protons from the reduction process without a scavenger will be incorporated in the cationic shells around the metal nanoparticles. This proton incorporation weakens the cationic shells so that surface-energy minimizing agglomeration can proceed to give larger particles and size distribution (Table 3.1.1 and Fig. 3.1.1).

The Ag-NP size obtained with BIm scavenger depends on the ionic liquid and increases roughly linear with the molecular volume of the IL anion (Table 3.1.1, entry 6-9 and Fig. 3.1.3). M-NPs (*core*) are considered stabilized in the ILs by the formation of "*protective*" anionic and cationic layers (shells) around them in a "*core-shell system*". We suggest that the thickness of the stabilizing shells around an Ag-NP depends on the IL molecular ion volumes. According to DLVO theory the first inner shell must be anionic, and then the IL anion will have the highest influence on the size and electrostatic stabilization of the Ag nanoparticle. The anion molecular volume determines the region of the nanoparticle size. Other physical IL parameters like density, viscosity or conductivity also correlate especially with the volume of the anion in the ionic liquid, although the supra-molecular imidazolium-anion clusters of the IL should be taken into account.

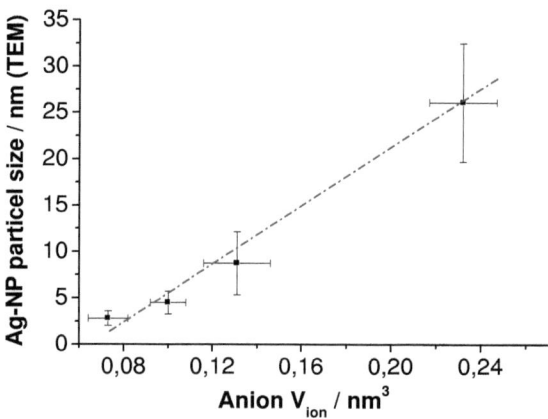

Figure 3.1.3: Correlation between the observed Ag nanoparticle size (from TEM) and the molecular volume ionic liquid anion ($V_{IL\text{-}anion}$).

Chapter 3.2: Use of Ionic Liquids (ILs) for the IL-Anion Size-Dependent Formation of Cr, Mo and W Nanoparticles from Metal Carbonyl M(CO)$_6$ Precursors

Cr, Mo and W-NPs were prepared by thermal or photochemical de-composition under argon of the mononuclear metal carbonyls M(CO)$_6$ in ILs (Scheme 3.2.1).

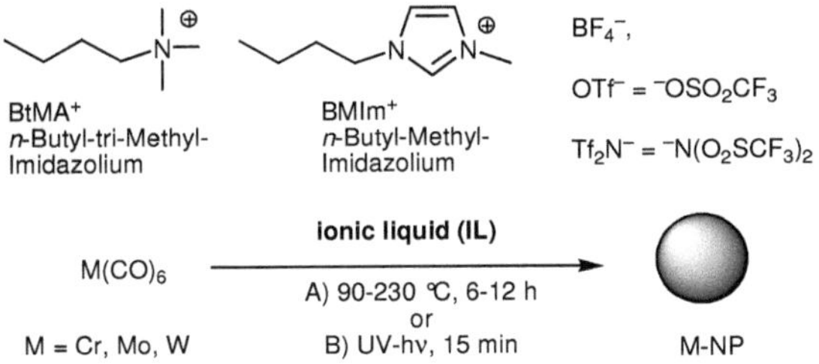

BtMA$^+$
n-Butyl-tri-Methyl-Imidazolium

BMIm$^+$
n-Butyl-Methyl-Imidazolium

BF$_4^-$,

OTf$^-$ = $^-$OSO$_2$CF$_3$

Tf$_2$N$^-$ = $^-$N(O$_2$SCF$_3$)$_2$

M(CO)$_6$

M = Cr, Mo, W

ionic liquid (IL)
A) 90-230 °C, 6-12 h
or
B) UV-hv, 15 min

M-NP

Scheme 3.2.1: Formation of Cr, Mo and W nanoparticles by thermal or photolytic decomposition of metal carbonyls under argon in ionic liquids

Green Cr, brown Mo or gray-blue W nanoparticle dispersions are obtained through decomposition from their metal carbonyl in ionic liquids. The dispersions are found to be stable for six months and longer under argon atmosphere. Also, extremely small M-NPs of Cr, Mo and W in the range of about 1 to 1.5 nm with a narrow, albeit not monodisperse, size distribution could be reproducibly synthesized in BMIm$^+$BF$_4^-$ (Fig. 3.2.1 and 3.2.2, Table 3.2.1). Nanoparticles of this small size are novel.

Nanoparticles obtained in BtMA$^+$Tf$_2$N$^-$ are larger, ranging from about 70 to 150 nm (Table 3.2.1). Such large nanoparticles can easily be separated, e.g. by simple centrifugation (10 min at 2000 rpm under argon) from the IL. TED (transmission electron diffraction) studies show that the larger Cr, Mo and W-NPs produced under argon are crystalline (Fig. 3.2.3), with the diffraction patterns corresponding to the metal lattices, thereby proofing their metallic character and the absence of significant oxidation.

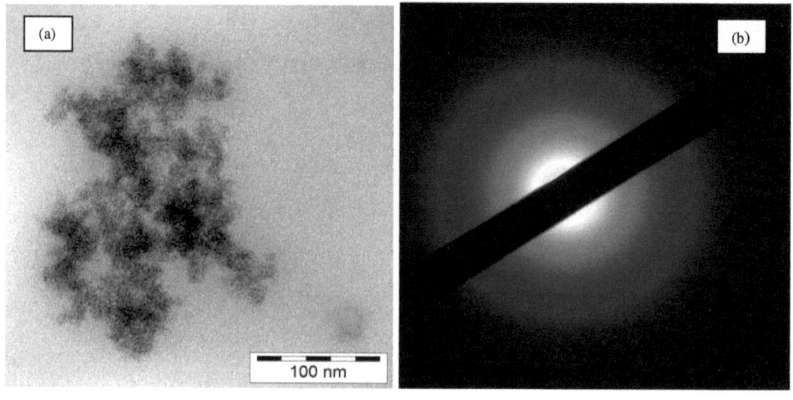

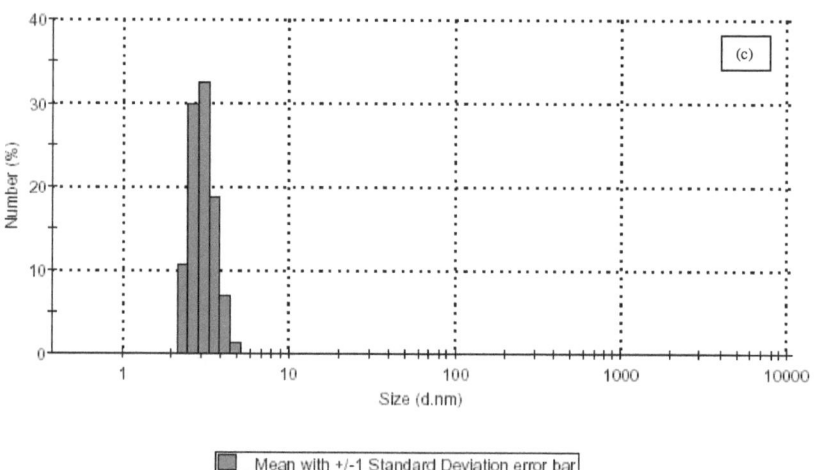

Fig. 3.2.1: (a) TEM of W-NPs from $W(CO)_6$ in $BMIm^+BF_4^-$ by thermal decomposition (entry 1 in Table 3.2.1); (b) TED picture of ultra small and amorphous W-NPs; (c) DLS (dynamic light scattering) measurement of W-NPs.

Furthermore, we synthesized the M-oxide nanoparticles for comparison to the Cr, Mo and W- NPs. For the M-oxides the $M(CO)_6$/IL mixture was subjected to the same decomposition conditions, albeit under air. The M-oxide NPs have different colors from the MNP suspensions, have a broader size distribution (Table 3.1.2) and show no crystallinity (Fig. 3.2.3).

A comparison of Mo-NPs produced under thermal und photolytic decomposition shows only slight differences in median size und size distribution (Table 3.2.1). Nanoparticles produced by photolysis give somewhat larger particles because of a faster decomposition process in the ionic liquid. (Fig. 3.2.4)

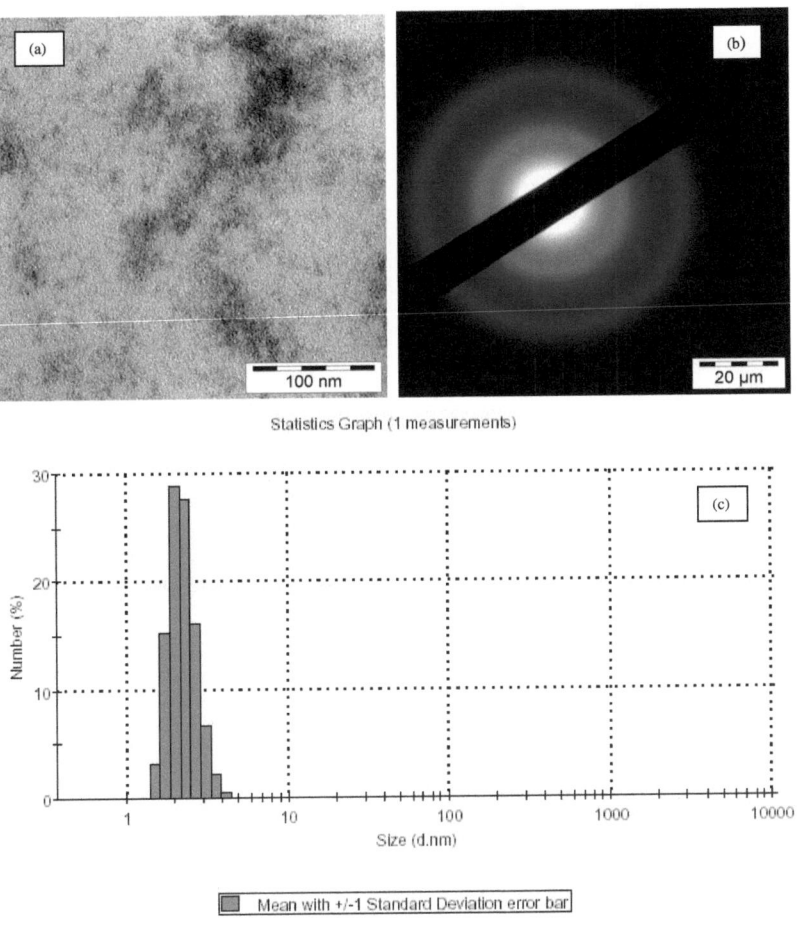

Fig. 3.2.2: (a) TEM of Mo-NPs from Mo(CO)$_6$ in BMIm$^+$BF$_4^-$ by thermal decomposition (entry 6 in Table 3.2.1); (b) TED picture of ultra small and amorphous Mo-NPs; (c) DLS (dynamic light scattering) measurement of Mo-NPs.

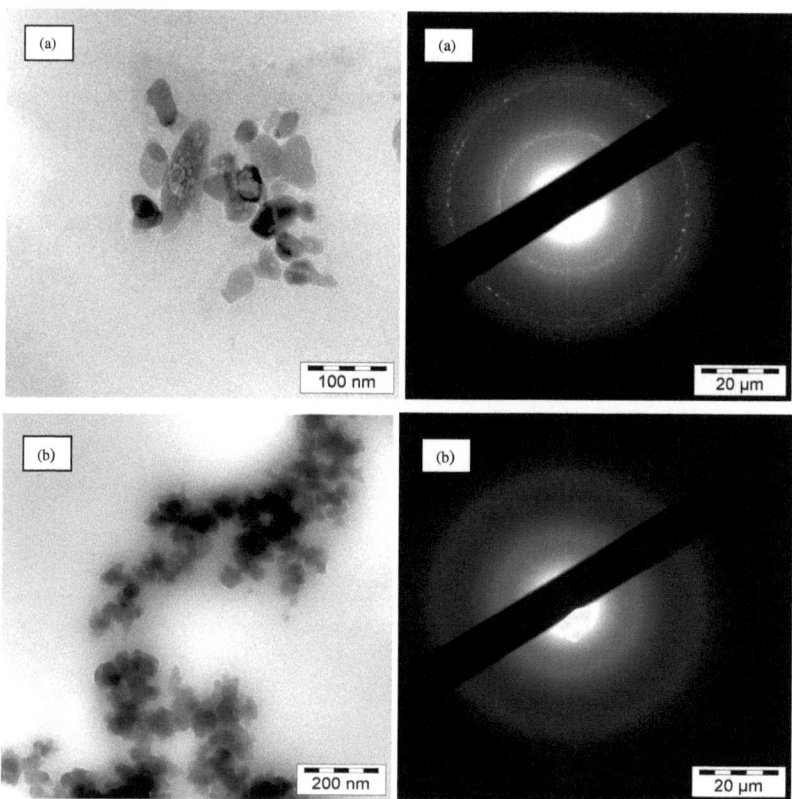

Fig. 3.2.3: (a) TEM/TED of big and crystalline W-NPs from $W(CO)_6$ in $BtMA^+Tf_2N^-$ by thermal decomposition (entry 4 in Table 3.2.1); The diffraction rings at (Å) 2.3 (very strong), 1.3 (strong), 1.6 (medium), 1.1 and 1.0 (all weak) match with D spacing of the W diffraction pattern.[173] (b) TEM/TED picture of big and amorphous Cr_2O_3 NPs.

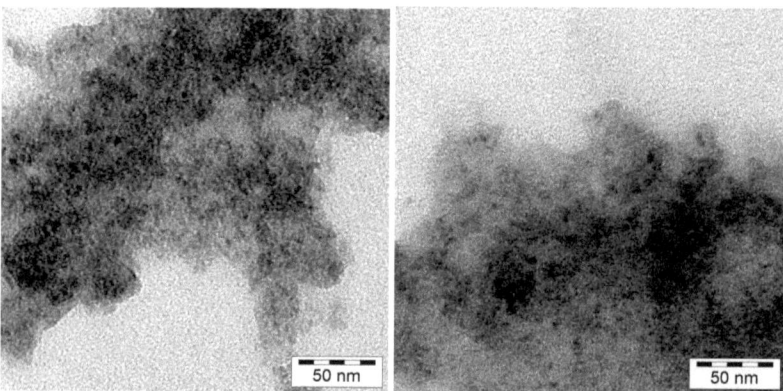

Fig. 3.2.4: TEM photograph of small Mo-NPs in $BMIm^+BF_4^-$ by photolytic decomposition.

Table 3.2.1: M-NP and M-oxide NP (M = Cr, Mo, W) size and size distribution in different ILs.

	Ionic liquid [a]	$V_{IL\text{-anion}}$ / nm^3 [185] (standard deviation σ)	Metal carbonyl	TEM NP median diameter / nm, (standard deviation σ) [b]	dynamic light scattering NP median diameter / nm, (standard deviation σ) [c]
1	BMIm$^+$BF$_4^-$	0.073 (± 0.009)	W(CO)$_6$	≤ 1.5 (± 0.3) [e]	3.1 (± 0.5)
2	BMIm$^+$OTf$^-$	0.131 (± 0.015)	W(CO)$_6$	5.7 (± 2.1)	11.7 (± 2.3)
3	BMIm$^+$ Tf$_2$N$^-$	0.232 (± 0.015)	W(CO)$_6$	33 (± 11)	45 (± 11)
4	BtMA$^+$Tf$_2$N$^-$	0.232 (± 0.015)	W(CO)$_6$	67 (± 32)	97 (± 33)
5	BtMA$^+$Tf$_2$N$^-$	0.232 (± 0.015)	W(CO)$_6$ / air	91 (± 83)	163 (± 47)
6	BMIm$^+$BF$_4^-$	0.073 (± 0.009)	Mo(CO)$_6$	≤ 1.5 (± 0.3) [e]	2.5 (± 0.6)
7	BMIm$^+$BF$_4^-$	0.073 (± 0.009)	Mo(CO)$_6$ [d]	~ 1.0 – 2.0 (± 0.6) [e]	3.8 (± 1.1)
8	BtMA$^+$Tf$_2$N$^-$	0.232 (± 0.015)	Mo(CO)$_6$	150 (± 30)	258 (± 89)
9	BtMA$^+$Tf$_2$N$^-$	0.232 (± 0.015)	Mo(CO)$_6$ / air	--- (layers, S19)	--- (layers)
10	BMIm$^+$BF$_4^-$	0.073 (± 0.009)	Cr(CO)$_6$	≤ 1.5 (± 0.3) [e]	3.0 (± 0.6)
11	BtMA$^+$Tf$_2$N$^-$	0.232 (± 0.015)	Cr(CO)$_6$	---	51 (± 12)
12	BtMA$^+$Tf$_2$N$^-$	0.232 (± 0.015)	Cr(CO)$_6$ / air	33 (± 10)	62 (± 16)

[a] BMIm$^+$ = n-butyl-methyl-imidazolium, BtMA$^+$ = n-butyl-trimethyl-ammonium [b] from TEM measurements, statistical evaluation of the total sample pictures. [c] hydrodynamic radius, median diameter form the first 3 measurements at 633 nm in n-butyl-imidazol. [d] photolytic decomposition 15 min at 200 to 450 nm. [e] The TEM pictures with particles of average median diameters of less than 1.5 nm show electron dense cloudy structures. Due to scattering caused by the surrounding ionic liquid resolution of the TEM is limited and particles below 1.5 nm are hardly resolved.

A correlation exists between the molecular volume of the anion in the ionic liquid and the synthesized metal nanoparticles, shown here for, but not limited to tungsten. The size of the Cr, Mo and W nanoparticles increases with the molecular volume of the IL anion (Table 3.2.1 and Fig. 3.2.5).

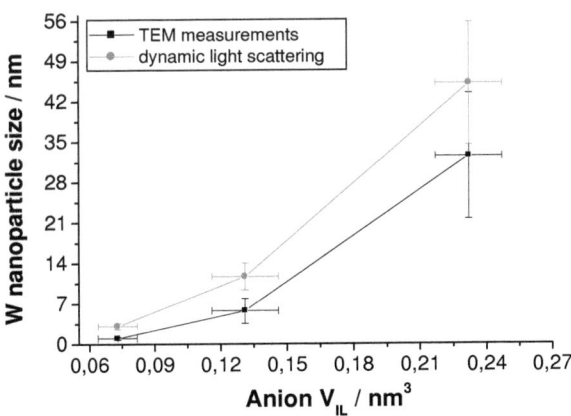

Fig. 3.2.5: Correlation between the molecular volume of the ionic liquid anion (V_{IL}) and the observed W nanoparticle size with standard deviations as error bars (from TEM and dynamic light scattering)

Chapter 3.3: Use of Ionic Liquids for the Synthesis of Fe, Ru and Os Nanoparticles from their Metal Carbonyl Precursors

Fe, Ru and Os metal nanoparticles were prepared by decomposition of di- and trinuclear metal carbonyls $Fe_2(CO)_9$, $Ru_3(CO)_{12}$ and $Os_3(CO)_{12}$, respectively, in n-butyl-methyl-imidazolium tetrafluoroborate, $BMIm^+BF_4^-$. In a typical experiment the metal carbonyl was dissolved/suspended under argon atmosphere in the dried and deoxygenated $BMIm^+BF_4^-$. For the M-NP synthesis the mixture was heated under argon up to 250 °C for several hours to thermally decompose the metal carbonyl. Alternatively, the mixture was irradiated at 200-450 nm for 15 min for photolytic decomposition (Scheme 3.3.1).

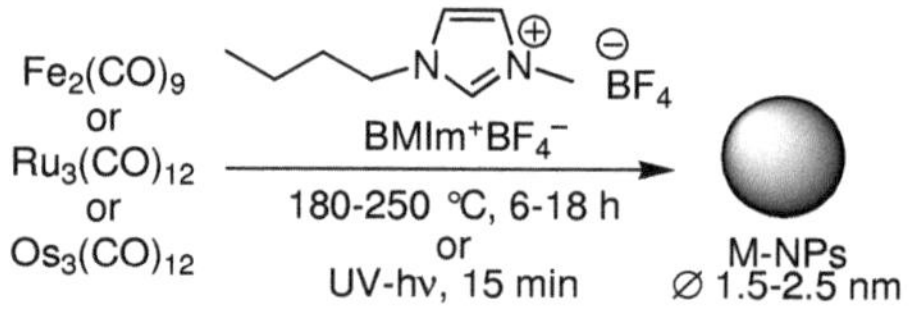

$Fe_2(CO)_9$ or $Ru_3(CO)_{12}$ or $Os_3(CO)_{12}$ $\xrightarrow[\text{or UV-hv, 15 min}]{BMIm^+BF_4^-,\ 180\text{-}250\ °C,\ 6\text{-}18\ h}$ M-NPs $\varnothing$ 1.5-2.5 nm

Scheme 3.3.1: Formation of Fe, Ru and Os nanoparticles by thermal and photolytic decomposition of metal carbonyls under argon in $BMIm^+BF_4^-$

Black Fe, dark-brown Ru or orange-red Os nanoparticle dispersions were reproducibly obtained and are stable for several months under argon. The median metal nanoparticle size for Ru and Os of 1.5 to 2.5 nm is extremely small with a narrow or uniform, albeit not monodisperse size distribution (Fig. 3.3.1). It is, at present, not trivial to routinely and easily prepare uniform nanoparticles of such small 1-2 nm size. No extra stabilizers or capping molecules are needed to achieve this small particle size. Fe nanoparticles are magnetic and agglomerate as a result of their superparamagnetic properties (Fig. 3.3.3).[150] Ru nanoparticles produced by thermal and photolytic decomposition show only slight differences in median size und size distribution (Fig. 3.3.2 and Table 3.3.1). Ru nanoparticles produced by photolysis give somewhat larger particles because of a faster decomposition and growth process in the ionic liquid.

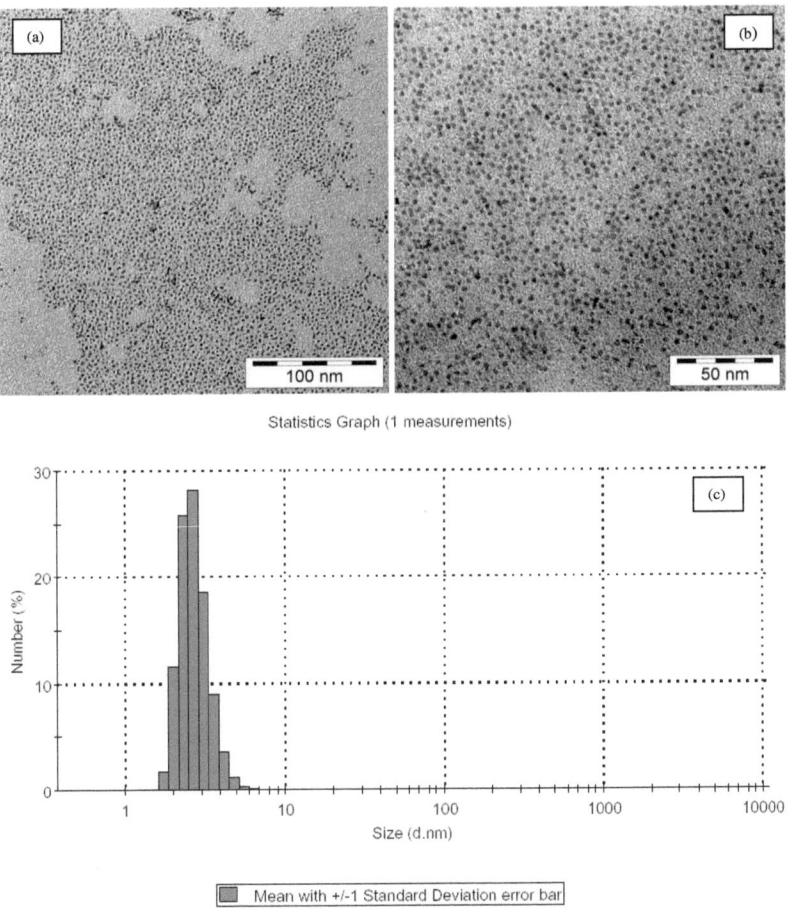

Fig. 3.3.1: TEM pictures of (a) Ru and (b) Os-NPs from $Ru_3(CO)_{12}$ and $Os_3(CO)_{12}$ in $BMIm^+BF_4^-$ by thermal decomposition (entry 5,9 in Table 3.3.1). (c) DLS (dynamic light scattering) measurement of Ru-NPs.

Moreover, we suggest that ionic liquids act as a "*novel nanosynthetic template*". The particle size does not significantly change with a five-fold difference in concentration of the precursor (0.2 vs. 1 wt. % Table 3.3.1). In nanoparticle syntheses through sol-gel, micro emulsion and other processes with stabilizers or capping molecules the concentration of the precursor plays a crucial role in determining the particle size and size distribution.[150,160] The nanoparticles can be separated by centrifugation (10 min at 2000 rpm under argon) from the ionic liquid and the ionic liquid can be reused.[174]

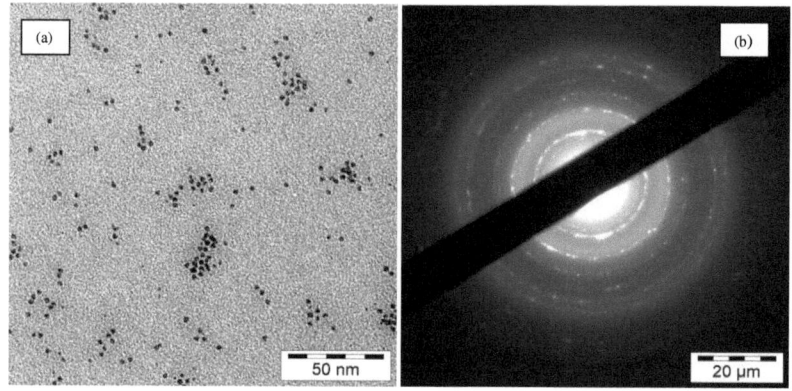

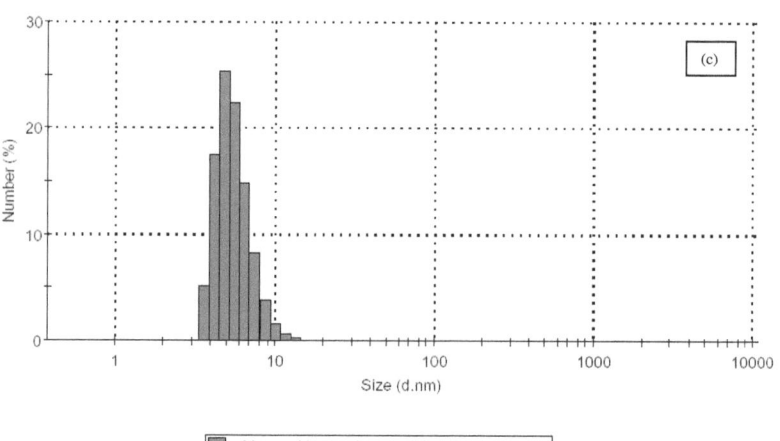

Fig. 3.3.2: (a) TEM picture of Ru-NPs, (b) TED of crystalline Ru-NPs from $Ru_3(CO)_{12}$ in $BMIm^+BF_4^-$ by photolytic decomposition (entry 8 in Table 3.3.1). (The black bar is the beam stopper.) The diffraction rings at (Å) 2.1 (very strong), 1.6 and 1.4 (strong), 1.2 and 1.1 (weak) match with D spacing of the Ru diffraction pattern.[173] (c) DLS (dynamic light scattering) measurement of Ru-NPs.

Furthermore, we synthesized Fe_2O_3 nanoparticles (Fig. 3.3.3) in air as a comparison and reference to the Fe-NP synthesis, so as to insure the absence of significant oxidation in the latter. For Fe_2O_3 the $Fe_2(CO)_9$/IL mixture was subjected to the same decomposition conditions, albeit under air. The rusty colored Fe_2O_3-NPs are not magnetic at room temperature, different to the black Fe-NPs. The Fe_2O_3-NPs are obtained as crystalline nanomaterials, based on electron diffraction analysis, again different to Fe nanoparticles which are amorphous under the same synthesis conditions (Fig. 3.3.3).

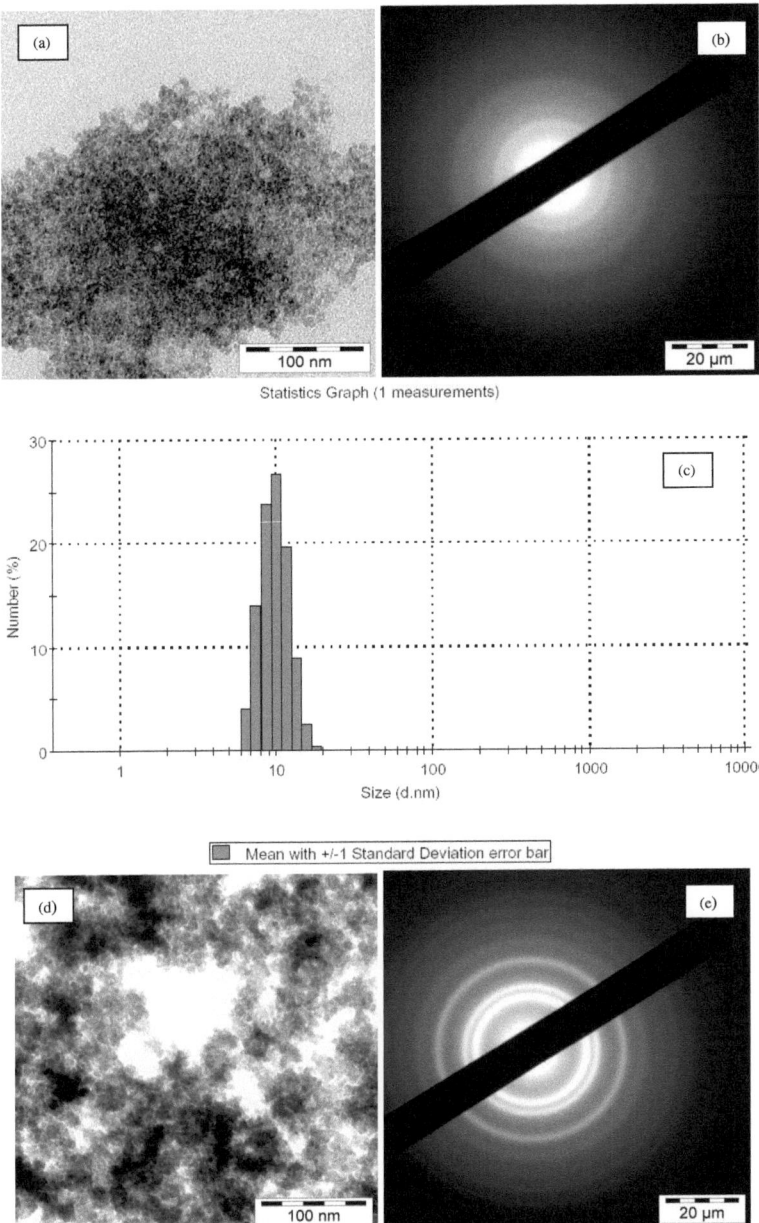

Fig. 3.3.3: TEM/TED of (a) TEM and (b) TED of amorphous Fe NPs from $Fe_2(CO)_9$ and (c) DLS (dynamic light scattering) measurement of Fe-NPs; (d) TEM of Fe_2O_3-NPs in $BMim^+BF_4^-$ by thermal decomposition (entry 2,3 in Table 3.3.1). (e) TED of crystalline Fe_2O_3-NPs (The black bar is the beam stopper.) The diffraction rings at (Å) 2.7 (very strong), 2.2 (strong), 1.6 (strong), 1.3, 1.1 and 1.0 (all weak) match with D spacing of the Fe_2O_3 (iron oxide) diffraction pattern.[173]

Table 3.3.1: Nanoparticle size and distribution in BMIm$^+$BF$_4^-$ analyzed by TEM and dynamic light scattering (DLS).

	Metal carbonyl / weight percent in IL [a]	Product [b]	TEM median diameter / nm (standard deviation σ) [c]	dynamic light scattering median diameter / nm (σ) [d]
1	Fe$_2$(CO)$_9$ / 0.2	Fe$_2$O$_3$	---	6.2 (± 1.2)
2	Fe$_2$(CO)$_9$ / 1	Fe$_2$O$_3$	4.2 (± 1.1)	6.4 (± 1.1)
3	Fe$_2$(CO)$_9$ / 0.2	Fe	5.2 (± 1.6)	10.1 (± 2.1)
4	Fe$_2$(CO)$_9$ / 1	Fe	---	10.7 (± 2.4)
5	Ru$_3$(CO)$_{12}$ / 0.2	Ru	1.6 (± 0.4)	2.9 (± 0.5)
6	Ru$_3$(CO)$_{12}$ / 0.6	Ru	---	2.9 (± 0.6)
7	Ru$_3$(CO)$_{12}$ / 1	Ru	---	2.8 (± 0.6)
8[e]	Ru$_3$(CO)$_{12}$ / 0.08	Ru	2.0 (± 0.5)	3.9 (± 1.0)
9	Os$_3$(CO)$_{12}$ / 0.2	Os	2.5 (± 0.4)	---
10	Os$_3$(CO)$_{12}$ / 1	Os	---	5.6 (± 1.5)

[a] Solubility of metal carbonyl precursors in BMIm$^+$BF$_4^-$ is limited to a maximum value of about 1 wt. %. [b] Fe$_2$O$_3$ assignment/analysis through electron diffraction. [c] Statistical evaluation of the total sample pictures. Transmission electron microscopy (TEM) photographs were taken at room temperature from a carbon coated copper grid on a Zeiss LEO 912 transmission electron microscope operating at an accelerating voltage of 120 kV. [d] Hydrodynamic radius, median diameter form the first 3 measurements at 633 nm. [e] Photolytic decomposition.

The obtained Ru nanoparticle/IL dispersions were tested by Dipl.-Chem. Christian Vollmer during his Diploma work for their catalytic activity in the biphasic liquid-liquid hydrogenation of cyclohexene to cyclohexane (Scheme 3.3.2).

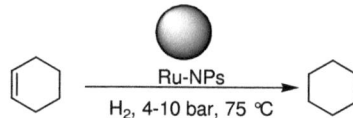

Scheme 3.3.2. Hydrogenation of cyclohexene catalyzed by Ru nanoparticles in BMIm$^+$BF$_4^-$.

Chapter 3.4: Synthesis of Co, Rh and Ir NPs from Metal Carbonyls in ILs and their use as Biphasic Liquid-Liquid Hydrogenation Nanocatalysts for Cyclohexen

Co, Rh and Ir metal nanoparticles were prepared by thermal decomposition of $Co_2(CO)_8$, $Rh_6(CO)_{16}$ and $Ir_4(CO)_{12}$, respectively, in different ILs. Moreover, the synthesized nanoparticles in ILs were used as highly effective and recyclable catalysts for the biphasic liquid-liquid hydrogenation of cyclohexene to cyclohexane.

In a typical experiment the metal carbonyl was dissolved under argon atmosphere in the dried and deoxygenated ionic liquid (IL). The solution was heated under argon up to 230 °C (above the decomposition temperature of the metal carbonyl) for several hours in the ionic liquids n-butyl-tri-methylammonium N-bis(trifluoromethylsulfonyl)imide ($BtMA^+NTf_2^-$), n-butyl-methyl-imidazolium tetrafluoroborate ($BMIm^+BF_4^-$) or trifluoromethanesulfonate ($BMIm^+OTf^-$) (Scheme 3.4.1). $Co_2(CO)_8$ was decomposed at 180 °C, $Rh_6(CO)_{16}$ and $Ir_4(CO)_{12}$ at 230 °C. Literature reported decomposition temperatures are 220 °C for $Rh_6(CO)_{16}$, 210 °C for $Ir_4(CO)_{12}$ and above 100 °C for $Co_2(CO)_8$.[175]

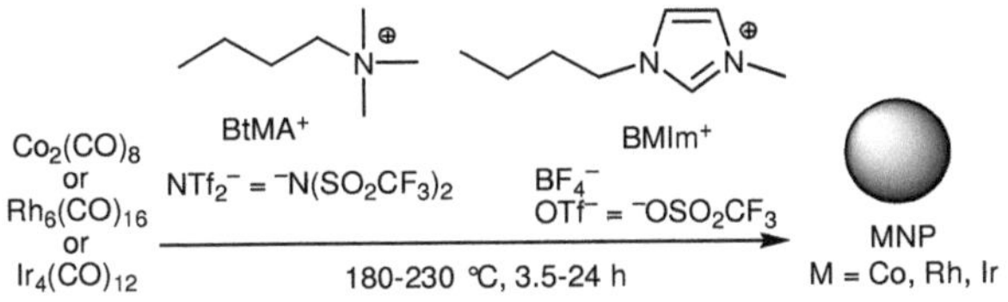

Scheme 3.4.1: Formation of Co, Rh and Ir nanoparticles by thermal decomposition of their metal carbonyls in different ionic liquids (ILs).

Dark-brown to black Co, Rh or Ir nanoparticle dispersions are reproducibly obtained through decomposition from their metal carbonyl in ionic liquids (Fig. 3.4.1-3.4.4). The resulting nanoparticles were analysed by transmission electron microscopy (TEM), transmission electron diffraction (TED) X-ray powder diffractometry (XRPD) and dynamic light scattering (DLS) (Table 3.4.1). The dispersions are found stable for more than six months under argon atmosphere. Under air the black, magnetic Co dispersion turns violet and loses its magnetic properties. TED und XRPD studies show that the resulting Co, Rh and Ir nanoparticles produced under argon are not oxidized. The XRPD diffraction patterns correspond to their metal lattices, e.g. for Co and Rh-NPs see Fig. 3.4.6, thereby proving their metal character and the absence of significant oxidation.[173]

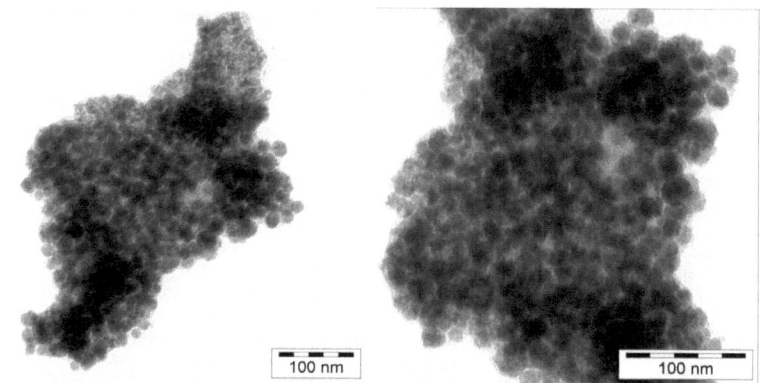

Fig. 3.4.1: Co NPs from thermal decomposition of $Co_2(CO)_8$ in $BMIm^+BF_4^-$ (entry 1 in Table 3.4.1).

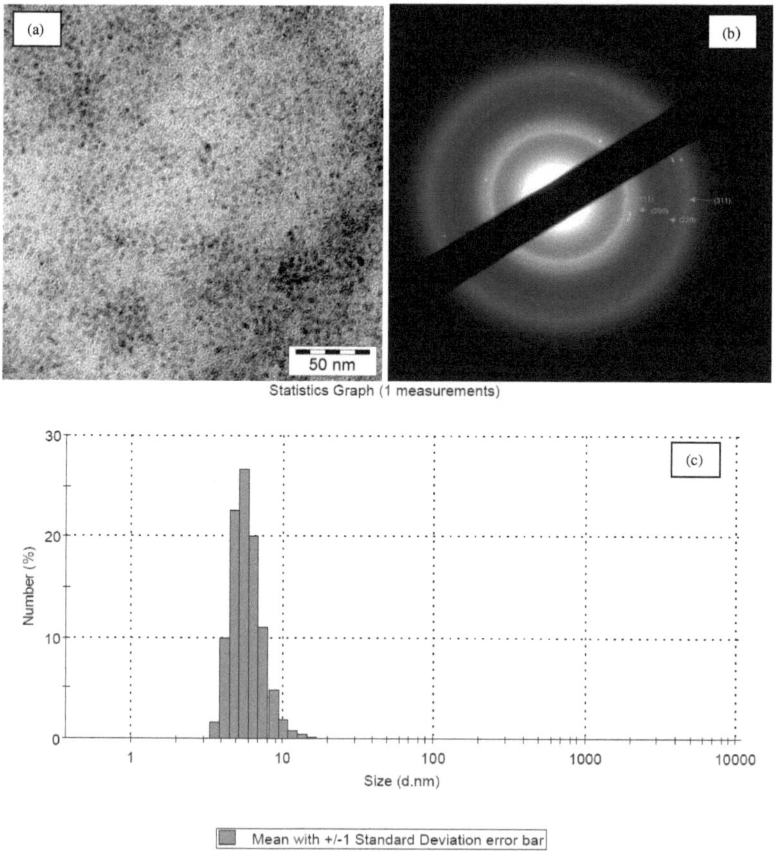

Fig. 3.4.2: (a) TEM of crystalline Rh-NPs from $Rh_6(CO)_{16}$ in $BMIm^+BF_4^-$ by thermolytic decomposition (entry 2 in Table 3.4.1). (b) TED of the Rh-NPs. (The black bar is the beam stopper.) The diffraction rings at (Å) 2.2 (very strong), 1.9 (strong), 1.3 and 1.1 (all weak) match with D spacing of the Rh diffraction pattern.[173] (c) DLS (dynamic light scattering) measurement of Rh-NPs.

The median metal nanoparticle size of ~ 3 nm for Rh and ~ 1 nm for Ir is extremely small with a narrow or uniform, albeit not monodisperse, size distribution (Fig. 3.4.2 and 3.4.3). It is, at present, not trivial to routinely and easily prepare uniform nanoparticles of such small 1-3 nm size. No extra stabilizers or capping molecules are needed to achieve this small particle size. The size of the Co, Rh and Ir nanoparticles in $BMIm^+BF_4^-$ could also be estimated with the Scherrer equation [176] from the half-width of the diffraction peaks.

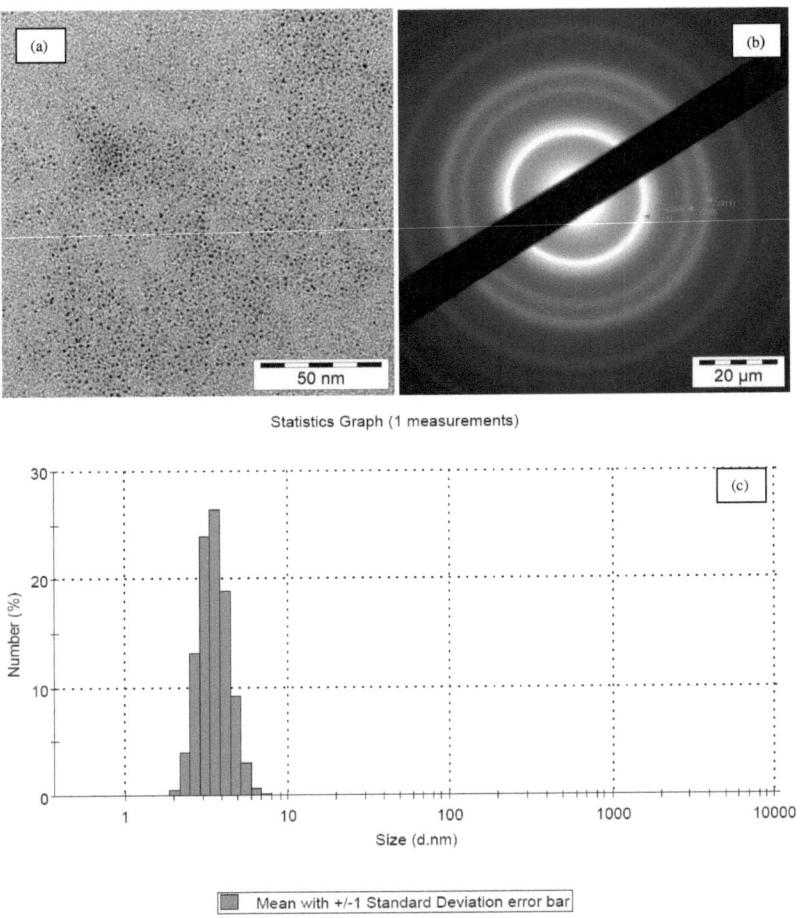

Fig. 3.4.3: (a) TEM of crystalline Ir-NPs from $Ir_4(CO)_{12}$ in $BMIm^+BF_4^-$ by thermolytic decomposition after 18 h (entry 7 in Table 3.4.1). (b) TED of the Ir-NPs. (The black bar is the beam stopper.) The diffraction rings at (Å) 2.2 (very strong), 1.9 (strong), 1.4 and 1.2 (all weak) match with D spacing of the Ir diffraction pattern.[173] (c) DLS (dynamic light scattering) measurement of Ir-NPs.

Ir nanoparticles obtained in BMIm⁺OTf⁻ and BtMA⁺NTf₂⁻ range from about ~4 nm in the former to ~80 nm the later (ableit this value corresponds to large aggregates of smaller Ir nanoparticles) (Fig. 3.4.4). Co nanoparticles obtained in BMIm⁺BF₄⁻ with a size of ~14 nm are magnetic and agglomerate as a result of their superparamagnetic properties. (Fig. 3.4.1)[150] The same observation of agglomerated Co nanoparticles was reported from a recent $Co_2(CO)_8$ decomposition in different imidazolium ionic liquid/hexane mixture, e.g., BMim⁺NTf₂⁻/hexane.[177]

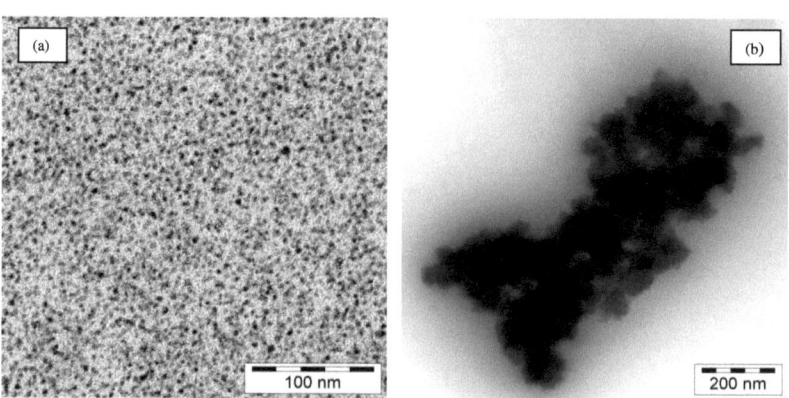

Fig. 3.4.4: (a) Ir nanoparticles from thermal decomposition of $Ir_4(CO)_{12}$ in BMim⁺OTf⁻ and (b) BtMA⁺NTf₂⁻ (entry 9 and 10, respectively, in Table 3.4.1).

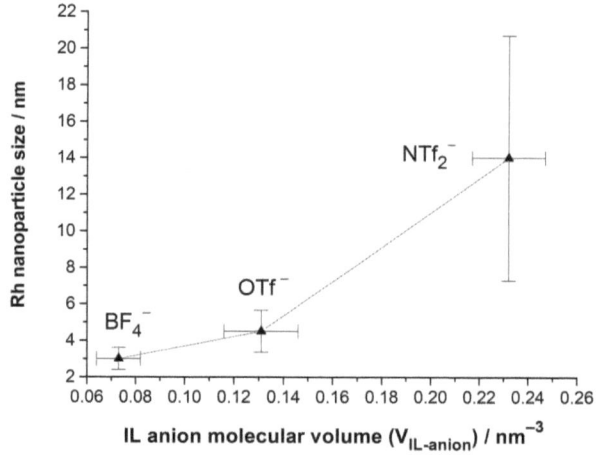

Fig. 3.4.5: Correlation between the molecular volume of the ionic liquid anion ($V_{IL\text{-anion}}$) and the observed Rh-NPs size with standard deviations as error bars (TEMs values, entry 2-4 in Table 3.4.1).

A correlation exists between the molecular volume of the anion in the ionic liquid and the synthesized metal nanoparticles,[58,60] indicated here for, but not limited to, Rh and Ir (Fig. 3.4.5). The crystallinity and the size of the Ir nanoparticles depend on the decomposition time. The longer the dispersions were heated, the more crystalline are the particles and the smaller their size (Table 3.4.1). The diffraction patterns of Ir samples which have been thermolyzed for 3.5 h show only diffuse halos. Ir nanoparticles from 18 h thermolysis show narrow and well defined reflections. (Fig. 3.4.3)

Table 3.4.1: M-NP (M = Co, Rh and Ir) size and size distribution in different ionic liquids (ILs).

	Ionic liquid [a]	$V_{\text{IL-Anion}}$ / nm^3 [178]	Metal carbonyl / metal weight percent in IL / decomposition time	TEM median diameter / nm (standard deviation σ) [b]	dynamic light scattering median diameter / nm (standard deviation σ) [c]
1	$BMIm^+BF_4^-$	0.073±0.009	$Co_2(CO)_8$ / 0.16 / 18 h	14 (± 8)	---
2	$BMIm^+BF_4^-$	0.073±0.009	$Rh_6(CO)_{16}$ / 0.2 / 18 h	3.0 (± 0.6)	6.0 (± 1.6)
3	$BMIm^+OTf^-$	0.131±0.015	$Rh_6(CO)_{16}$ / 0.2 / 18 h	4.4 (± 1.1)	9.5 (± 1.7)
4	$BtMA^+NTf_2^-$	0.232±0.015	$Rh_6(CO)_{16}$ / 0.5 / 18 h	14 (± 7)	---
5	$BMIm^+BF_4^-$	0.073±0.009	$Rh_6(CO)_{16}$ / 1 / 18 h [d]	3.5 (± 0.8)	7.0 (± 1.2)
6	$BMIm^+BF_4^-$	0.073±0.009	$Ir_4(CO)_{12}$ / 0.2 / 3.5 h	1.1 (± 0.2)	4.1 (± 0.7)
7	$BMIm^+BF_4^-$	0.073±0.009	$Ir_4(CO)_{12}$ / 0.2 / 18 h	1.7 (± 0.3)	3.6 (± 0.7)
8	$BMIm^+BF_4^-$	0.073±0.009	$Ir_4(CO)_{12}$ / 0.5 / 18 h	1.3 (± 0.2)	3.4 (± 1.0)
9	$BMIm^+OTf^-$	0.131±0.015	$Ir_4(CO)_{12}$ / 0.5 / 18 h	3.6 (± 0.6)	7.1 (± 1.1)
10	$BtMA^+NTf_2^-$	0.232±0.015	$Ir_4(CO)_{12}$ / 0.5 / 18 h	81 (± 23) [e]	---

[a] Solubility of metal carbonyl precursors in $BMIm^+BF_4^-$ is limited to a maximum value of about 1 wt%. – [b] Statistical evaluation of the total sample pictures. – [c] Hydrodynamic radius, median diameter form the first 3 measurements at 633 nm. The hydrodynamic radius is roughly 2-3 times the size of the pure kernel cluster. For very small MNPs (~1 nm) the size of the hydrodynamic radius can even increase to more than 3 times the MNP radius. The resolution of the DLS instrument is 0.6 nm. – [d] After six catalytic cycles. – [e] This value corresponds to large aggregates of smaller Ir nanoparticles.

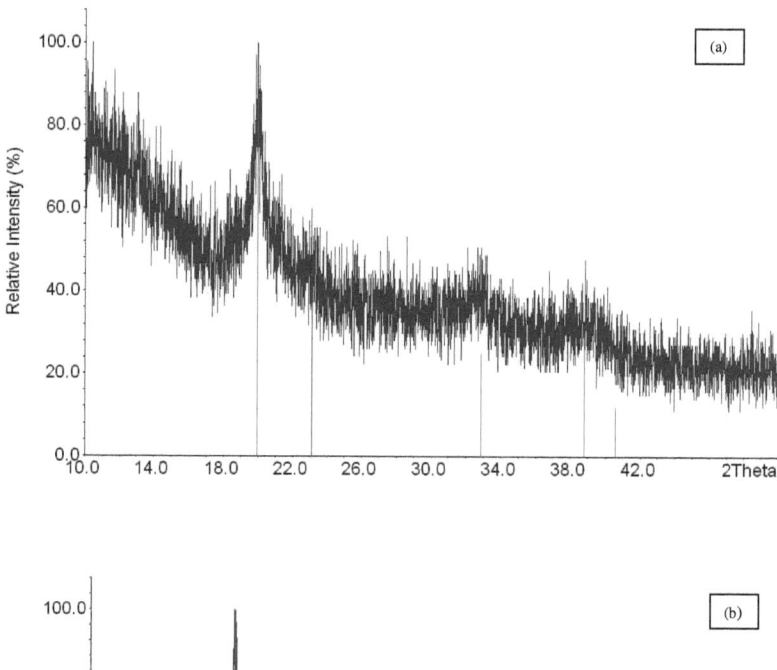

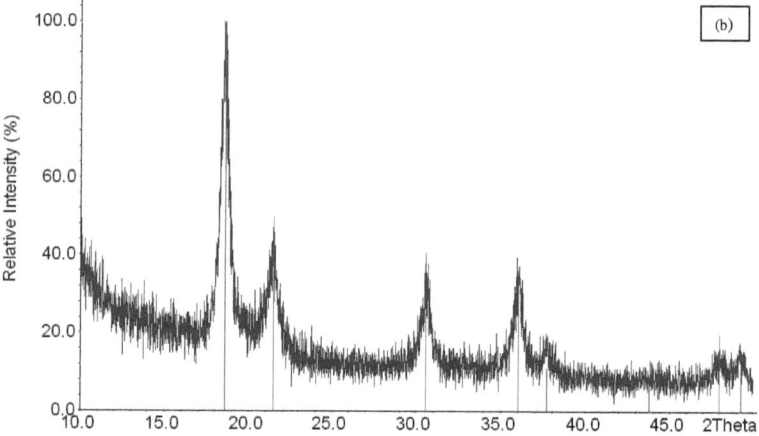

Fig. 3.4.6: (a) XRPD of Co-NPs compared with reflexes of Co^0 from the WinXPOW-Database [15-806]. Calculation Scherrer Equation: Ø 19.37 nm (b) XRPD of Rh-NPs compared with reflexes of Rh^0 from the WinXPOW-Database [5-685]. Calculation Scherrer Equation: Ø 3.74 nm

Cyclohexene Hydrogenation Catalysis with MNP/IL

The obtained Co, Rh and Ir nanoparticle/IL dispersions were tested by Dipl.-Chem. Jérôme Krämer during his Diploma work for their catalytic activity in the biphasic liquid-liquid hydrogenation of cyclohexene to cyclohexane (Scheme 3.4.2).

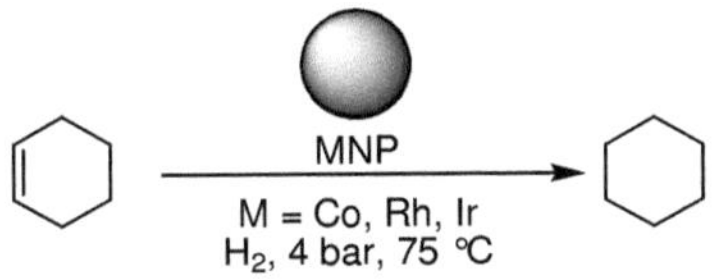

Scheme 3.4.2: Hydrogenation of cyclohexene catalyzed by Co, Rh and Ir nanoparticles in BMIm$^+$BF$_4^-$.

While the Co nanoparticles showed very little to no activity, Ir and Rh nanoparticles were found to be highly active nanocatalysts under very mild reaction conditions (75 °C, 4 bar H$_2$ pressure). Especially Ir nanoparticles exhibited very high catalytic activities of up to 1900 mol product · (mol metal)$^{-1}$ · h^{-1}. Rh nanoparticles showed good catalytic activities in the range of 300 to 380 mol product · (mol metal)$^{-1}$ · h^{-1} through different catalytic runs. The higher catalytic activity of Ir in comparison to Rh nanoparticles may be explained by their somewhat smaller size and concomitant lager surface-to-volume ratio (cf. Table 3.4.1).

ILs are once again demonstrated to present a favourable template for the preparation of predictable chemical nanostructures. Furthermore, the obtained Rh-NP and Ir-NP/IL dispersions were employed as highly active and reusable *"green catalysts ??"* without the need of other organic solvents in the biphasic liquid-liquid hydrogenation of cyclohexene.

Chapter 3.5: Synthesis, Stabilization, Functionalization and DFT Calculations of Gold Nanoparticles in Fluorous Phases (PTFE and ILs)

3.5.1 Au-NP synthesis

In a typical experiment the gold metal precursor was dissolved in the presence of *n*-butyl-imidazole under argon atmosphere in the carefully dried and deoxygenated ionic liquid (IL). The solution was heated under argon to 230 °C for several hours or treated by photolytic (hv) or microwave (MW) radiation for some minutes in the ionic liquids *n*-butyl-tri-methylammo-nium *N*-bis(trifluoromethylsulfonyl)imide (BtMA$^+$ NTf$_2^-$), *n*-butyl-methyl-imidazolium tetrafluoroborate (BMIm$^+$BF$_4^-$) or trifluoromethanesulfonate (BMIm$^+$OTf$^-$) (Scheme 3.5.1). The compound Au(CO)Cl[179] starts to decompose at 140 °C as evidenced by the gas evolution.

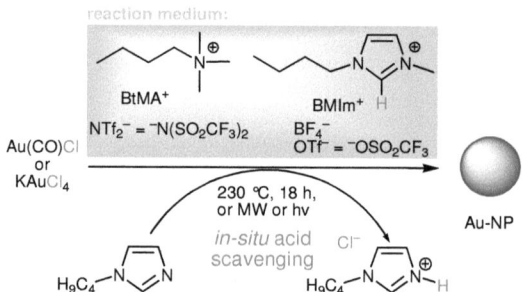

Scheme 3.5.1: Formation of Au nanoparticles by thermal decomposition and reduction of different metal precursors ILs. The formed HBIm$^+$Cl$^-$ was analyzed by elemental analysis and ^{1}H-NMR.

Yellow-orange, red or red-purple to brown Au-NP dispersions are reproducibly obtained through decomposition and reduction from their metal precursors in ILs (see, Fig. 3.5.1 and 3.5.2, Table 3.5.1). Upon exposure to air a yellow-orange Au dispersion in BMIm$^+$BF$_4^-$ forms a red-purple precipitate, indicating particle aggregation (Fig. 3.5.3). Without the presence of *n*-butyl-imidazole the Au-NP/IL dispersion also quickly turns red-purple, indicating an agglomeration process which is caused by the generated free HCl acid. As for Pd-NPs, we suggest here the formation of *N*-heterocyclic carbene (NHC) Au species as the origin of the formation of Au-NPs (Scheme 3.5.2).[180,181] We also suggest that the formation of these Au-NHC intermediates slows down the aggregation process. In the presence of *n*-butyl-imidazole the released HCl is bound as a new imidazolium salt, similar to the IL matrix. This prevents the formation of an acidic reaction medium which would destabilize the Au-NPs and lead to their clustering. The Au-NPs with a size of less than 2 nm are unstable when prepared without a scavenger in ILs. In the absence of the BIm scavenger, the Au-NP size is larger and their size distribution is broader (DLS ~4.5 and 15 to 35 nm). This can be reasoned by proton (H⁺/HCl) incorporation in the IL dynamic matrix which weakens its stabilizing effect so that larger nanoparticles and size distributions are obtained.[48] In addition, free 1-methyl-imidazole can stabilize Au-NPs as a ligand in ILs (see DFT calculation in Section 3.5.5).[182]

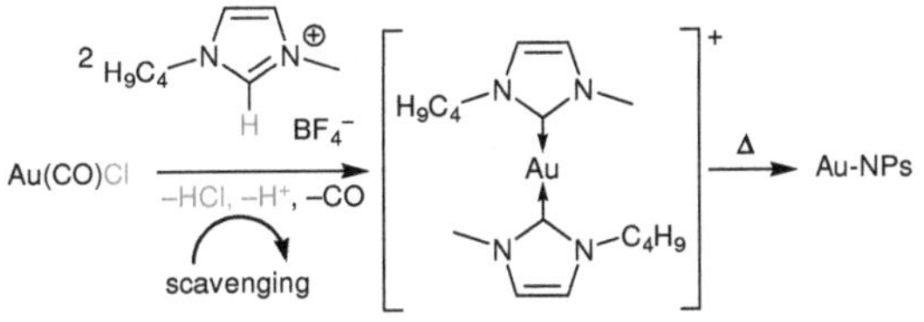

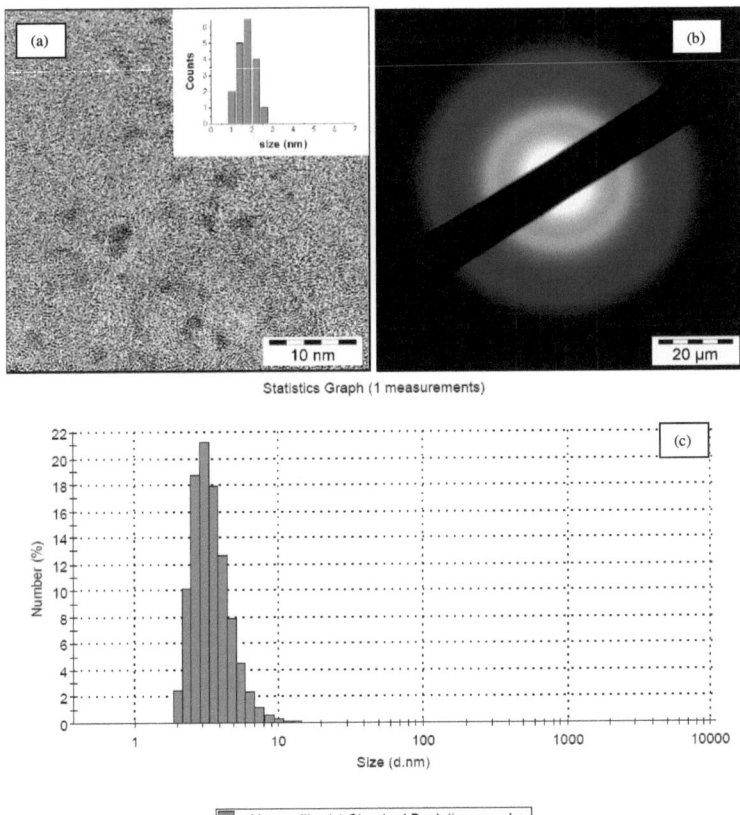

Scheme 3.5.2: Suggested Au-NP formation through the formation of an Au-carbene intermediate.[180,181]

Fig. 3.5.1: (a) HRTEM of Au-NPs from Au(CO)Cl in BMIm⁺BF₄⁻ by thermal decomposition (entry 1a in Table 3.5.1); (b) TED picture of small and amorphous Au-NPs. (The black bar is the beam stopper.) (c) DLS (dynamic light scattering) measurement of Au-NPs.

In BMIm$^+$BF$_4^-$ the average size of the gold nanoparticles from Au(CO)Cl or KAuCl$_4$ is about 1–2 nm with an extremely small and uniform, albeit not monodisperse, size distribution (Fig. 3.5.1 and 3.5.2). Currently, it is not trivial to routinely and easily prepare uniform Au nanoparticles of such small 1–2 nm size. No extra stabilizers or capping molecules are needed to achieve this small particle size. Furthermore, we observed that the ultra small nanoparticles (Table 3.5.1) agglomerate in a controlled way by the fusion of two Au-NPs with time and faster when they are exposed to air (Fig. 3.5.3). Thereby, the particle size doubles (see Fig. 3.5.1 and Fig. 3.5.3, entry 1b and c in Table 3.5.1) with a still quite uniform size distribution

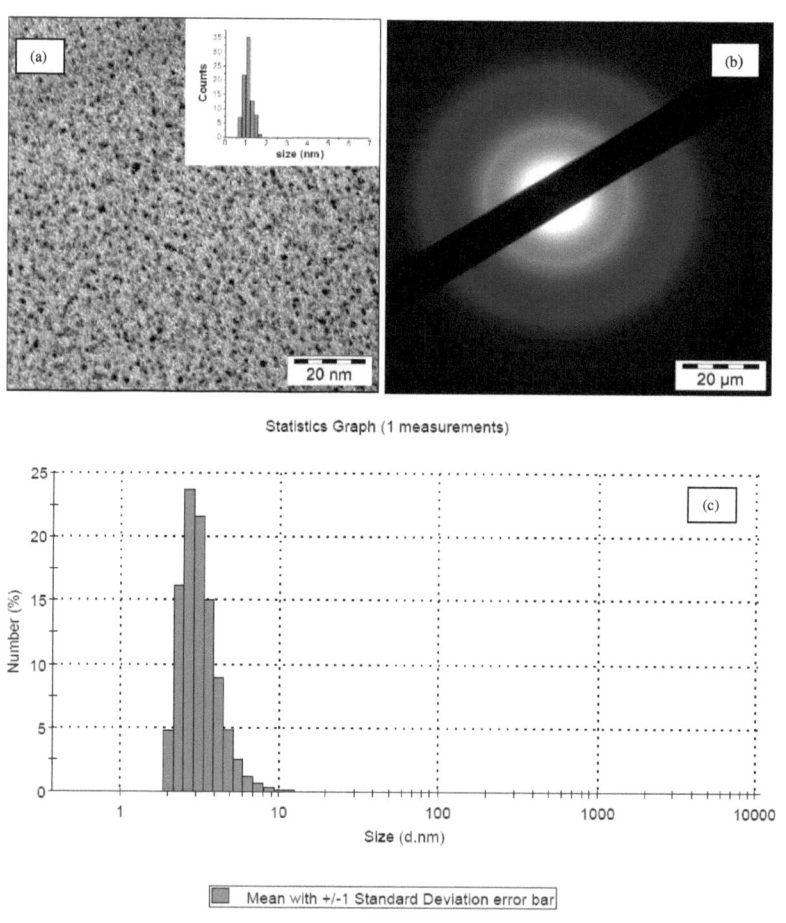

Fig. 3.5.2: (a) TEM of Au-NPs from KAuCl$_4$ in BMIm$^+$BF$_4^-$ by thermal/reduction decomposition (entry 2 in Table 3.5.1); (b) TED picture of small and amorphous Au-NPs. (The black bar is the beam stopper.) (c) DLS (dynamic light scattering) measurement of Au-NPs.

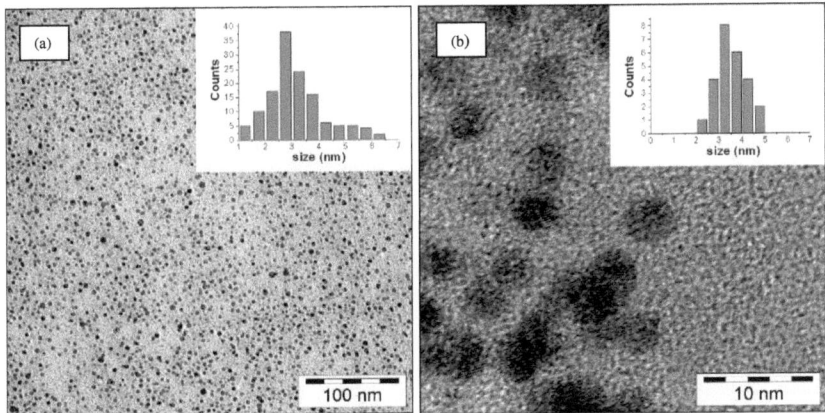

Fig. 3.5.3: (a) Au-NP precipitate from Au(CO)Cl in BMIm⁺BF₄⁻ after two weeks under argon (TEM, left, entry 1b in Table 3.5.1) and (b) after 6 weeks of expose to air (HRTEM, right, entry 1c in Table 3.5.1).

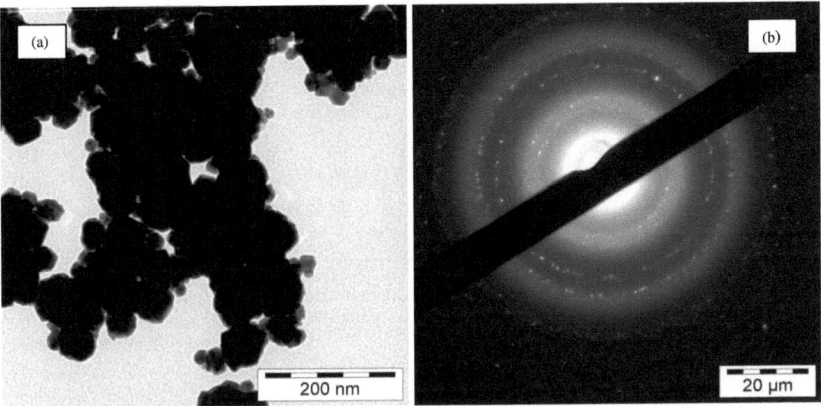

Fig. 3.5.4: (a) Au-NP from Au(CO)Cl in BMIm⁺OTf⁻ by thermal decomposition (TEM, left, entry 5 in Table 3.5.1) and (b) TED picture of big and crystalline Au-NPs. (The black bar is the beam stopper.) The diffraction rings at (Å) 2.4 (very strong), 2.1 (strong), 1.5 (medium), 1.3 and 1.0 (all weak) match with D spacing of the Au diffraction pattern.[173]

The use of BMIm⁺OTf⁻ and BtMA⁺NTf₂⁻ as ionic liquids results in much bigger Au nanoparticles with a broad size distribution of about 130 ± 40 nm and 350 ± 180 nm, respectively (see Fig. 3.5.4 and Table 3.5.1), compared to BMIm⁺BF₄⁻. We have previously shown that a larger NP size correlates with a larger volume of the ionic liquid anion. Agglomeration to larger Au nanoparticles in ionic liquids (ILs) is not prevented because of the only weak interaction of the IL species with the Au-NPs and, thus, is driven by the known strong aurophilic Au-Au interaction (Au cohesive energy 3.8 eV, ~88 kcal/mol).[183]

The use of strong heat sources in the microwave or photolytic decomposition (microwave, 10 W, 250 °C or light energy, 1000 W, see Fig. 3.5.5) accelerates the Au-NP agglomeration process in the IL network. Microwave or light energy leads to highly localized "*hot spots*" [184] through the energy absorption of the Au-NPs.

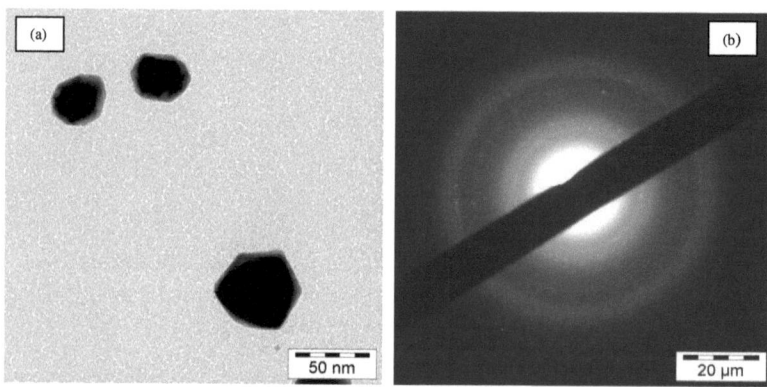

Fig. 3.5.5: a) Au-NP from Au(CO)Cl in $BMIm^+BF_4^-$ by photolytic decomposition (TEM, left, entry 5 in Table 3.5.1) and b) TED picture of medium and crystalline Au-NPs. (The black bar is the beam stopper.) The diffraction rings at (Å) 2.4 (very strong), 2.1 (strong), 1.5 (medium), 1.3 and 1.0 (all weak) match with D spacing of the Au diffraction pattern.[173]

Table 3.5.1: Au-NPs from Au(I) and Au(III) precursors with size and size distribution in different ionic liquids (ILs).

	Ionic liquid [a]	Metal precursor (1 wt% in IL) [a] / decomposition time	TEM & HRTEM median diameter / nm (standard deviation σ) [b]	dynamic light scattering median diameter / nm (standard deviation σ) [c]
1a	$BMIm^+BF_4^-$	Au(CO)Cl / 18 h	1.8 (± 0.4)	3.6 (± 1.3)
1b	$BMIm^+BF_4^-$	Au-NPs from 1a after 2 weeks under Ar	3.0 (± 0.9)	---
1c	$BMIm^+BF_4^-$	Au-NPs from 1a after 6 weeks under air	3.6 (± 0.6)	7.3 (± 1.4)
2	$BMIm^+BF_4^-$	$KAuCl_4$ / 18 h	1.1 (± 0.2)	3.3 (± 1.1)
3	$BMIm^+BF_4^-$	Au(CO)Cl / MW 5min	4.1 (± 0.7)	8.0 (± 2.2)
4	$BMIm^+BF_4^-$	Au(CO)Cl / hv 3min	61 (± 43)	154 (± 76)
5	$BMIm^+OTf^-$	Au(CO)Cl / 18 h	130 (± 40)	280 (± 140)
6	$BtMA^+NTf_2^-$	Au(CO)Cl / 18 h	350 (± 180)	600 (± 180)

[a] Solubility of metal carbonyl precursors in $BMIm^+BF_4^-$ is limited to a maximum value of about 1-2 wt%. Anion volumes: BF_4^- 0.073±0.009 nm^3; OTf^- 0.131±0.015 nm^3; NTf_2^- 0.232±0.015 nm^3 [185] [b] Statistical evaluation of the total sample pictures [c] Hydrodynamic radius, median diameter from the first 3 measurements at 633 nm. The hydrodynamic radius is roughly 2-3 times the size of the pure kernel cluster. For very small Au NPs (~1 nm) the size of the hydrodynamic radius can even increase to more than 3 times the Au NP radius. The resolution of the DLS instrument is 0.6 nm.

Such activated particles will be prone to agglomeration. The diffraction patterns from transmission electron diffraction (TED) correspond to the Au metal lattices (Fig. 3.5.4 and 3.5.5) thereby proving their metal character and the absence of significant oxidation. The crystallinity increases with the size of the Au nanoparticles.[60]

Nanoparticles must be stabilized in order to prevent their agglomeration or aggregation which eventually leads to the formation of small bulk metal particles. Metal nanoparticles (M-NPs) (*core*) are considered stabilized in ILs by the formation of "*protective*" anionic and cationic layers (*shells*) around them in a "*core-shell system*".[186,187,188,189] Pure ILs can be considered as supramolecular polymeric structures with a high degree of self-organisation and weak interactions of an extended network of cations and anions connected together by hydrogen bonds.[6] When mixed with other molecules or M-NPs, ionic liquids become nano-structured materials with polar and nonpolar regions.[190,191,192,193] The ionic liquid represents an excellent dynamic nanoenvironment for the formation of M-NPs with, in most cases, a small size and narrow size distribution under mild reaction conditions. We suggest that weak fluorous Au$\cdots$F interaction from the IL anion to the Au nanoparticles may aid in the stabilization (see DFT calculation in Section 3.5.5). Also, such fluorous M$\cdots$F (M = electrophilic transition metals) anion interaction in ionic liquids are not limited to Au nanoparticles.

3.5.2 Au-NP surface functionalization

The addition of an organic ligand to the bare M-NP surface is generally described as a surface functionalization, albeit derivatization, coating or capping maybe better terms. The post synthetic introduction of an organic capping ligand on the dispersed gold nanoparticles in ILs was done by treating the gold dispersion with an excess of mercaptopropionic acid (HS-$(CH_2)_2$-COOH), thioglycolic acid (HS-CH_2-COOH), *n*-decanthiol (HS-$C_{10}H_{21}$), *n*-dodecanthiol (HS-$C_{12}H_{25}$), and *n*-tetradecanthiol (HS-$C_{14}H_{29}$), respectively, at room temperature (Scheme 3.5.3, Table 3.5.1). The strong affinity between the thiol (-SH) group and the gold nanoparticles, replaces the ionic liquid protective layer.

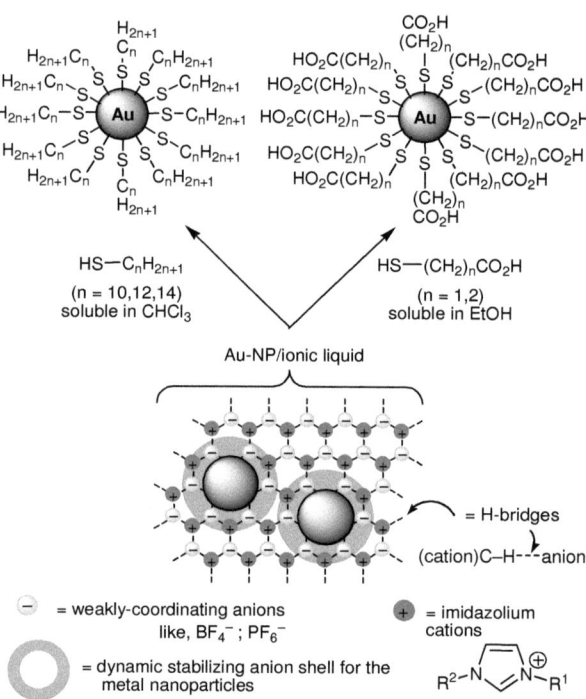

Scheme 3.5.3: Stabilization and surface capping (functionalization, derivatization) of Au nanoparticles from the ionic liquid network by hydrophobic and hydrophilic thiols ligands.

Surface capping of gold nanoparticles dispersed in the ionic liquid $BMIm^+BF_4^-$ with a polar and soluble organic thiol-ligand leads to *slightly* larger thiol-capped gold nanoparticles. (Fig. 3.5.7 and Table 3.5.2, entry 1-2). The use of a non-polar and insoluble organic thiol ligand more than doubles the size of the resulting thiol-capped gold nanoparticles (Fig. 3.5.6 and Table 3.5.2, entry 3-5). The aggregation is a result of the introduction of the thiol ligands into the ionic liquid network which disrupts their hydrogen-bonded network. Subsequently the stabilizing property of the ionic liquid network towards the Au-NPs is weakened and results in further Au-NP agglomeration which is driven by the aurophilic Au-Au interaction (see, Table 3.5.2). The interaction with the thiol ligand results in a strong Au-S chemisorption bond.[194] The addition of a thiol ligand leads to larger Au-NPs based on a controlled aggregation of the gold nanoparticles. The larger Au-thiol covered nanoparticles could be collected by centrifugation. The IL together with the excess thiol ligand is decanted and the Au/thiol-NPs were washed twice with a water-methanol solution followed by methanol.[195] The thiol-capped Au nanoparticles can then be redispersed in polar or non-polar organic solvents like ethanol or chloroform (Fig. 3.5.6 and 3.5.7).

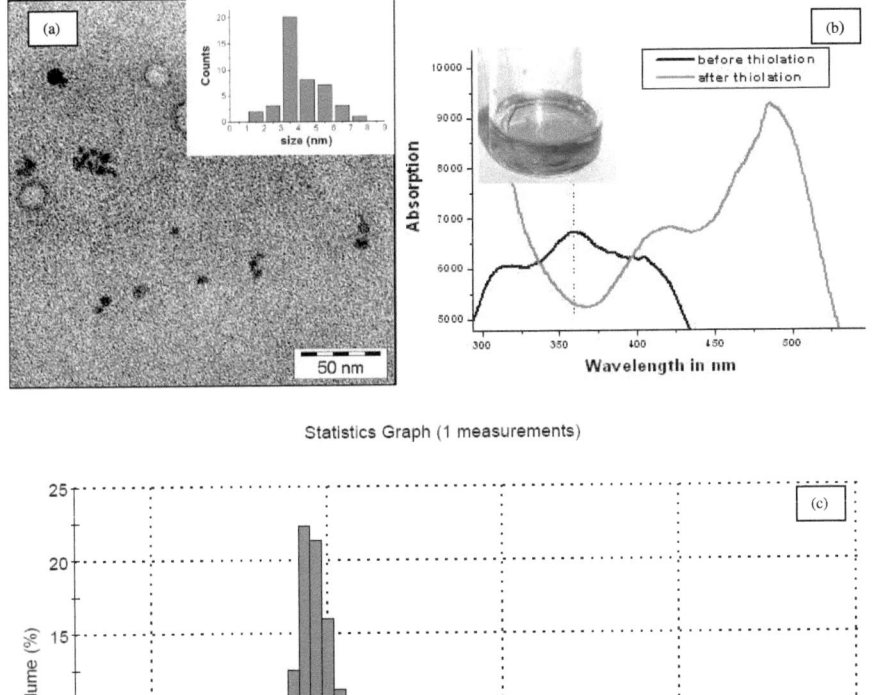

Fig. 3.5.6: (a) Au/SR-NPs from BMIm$^+$BF$_4^-$ after thiolation with decanthiol, centrifugation and redispersion in CHCl$_3$ TEM picture and (b) UV/VIS spectrum (right, red curve). (c) DLS (dynamic light scattering) measurement of Au-NPs.

The thiol-capped Au/SR-NP dispersions in ethanol and CHCl$_3$ are stable for months. The particle size does not change with time, even with air/oxgen contact. TED studies show that the thiol-capped Au nanoparticles are not oxidized. The diffraction patterns from TED correspond to their metal lattices[196] thereby proving their metallic character and the absence of significant oxidation.

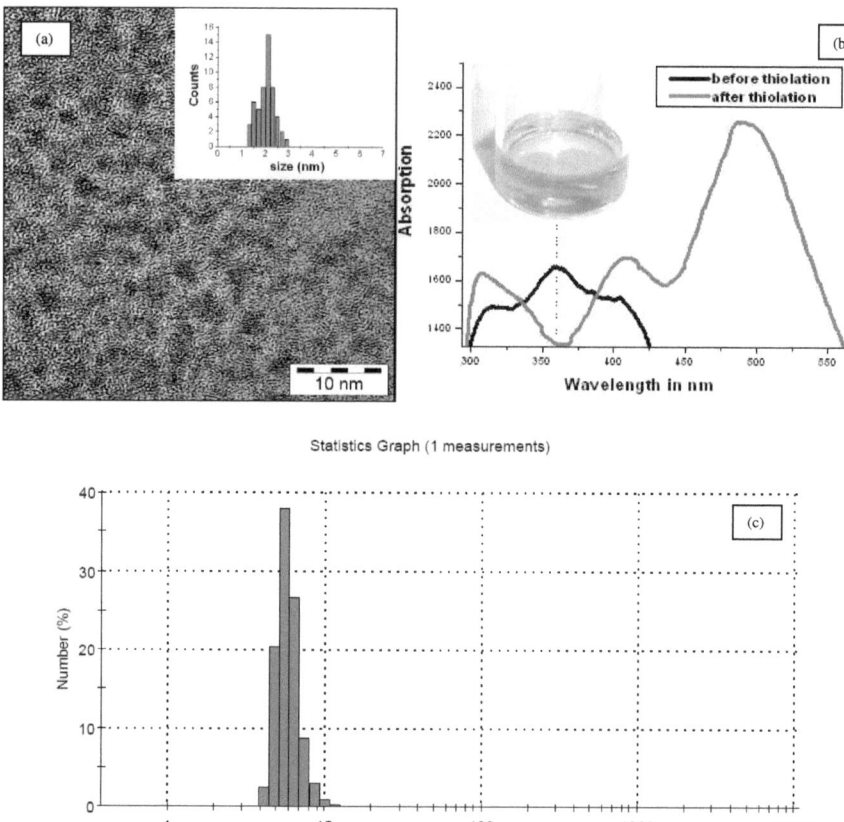

Fig. 3.5.7: (a) Au/SR-NPs from BMIm⁺BF₄⁻ after thiolation with mercaptopropionicacid, centri-fugation and redispersion in ethanole HRTEM picture and (b) UV/VIS spectrum (right, red curve). (c) DLS (dynamic light scattering) measurement of Au NPs.

Thiolation replaces the weakly-bound $BMIm^+BF_4^-$ coating from the dispersed Au-NPs thereby leading to a structural organisation of the Au clusters and to a controlled agglomeration process[197] (compare Fig. 3.5.1, 3.5.6 and 3.5.7, Table 3.5.2). Thereby, the process of thiol capping of gold nanoparticles in ionic liquids has similarity with a ligand exchange process. It is already known that the exchange of the stabilizers or capping molecules from the nanoparticles surface results in clustering of different defined gold species, e.g. from phosphane or chloride stabilized Au_{11} to thiol stabilized Au_{25}[198] and other size changes.[199] In particular for small clusters this can be understood due to the strong correlation of specific stable cluster sizes with the nature of a given ligand.[200,201]

Table 3.5.2: Thiol-capped Au/SR-NP from BMIm$^+$BF$_4^-$ with size distribution in ethanol and chloroform.

	Organic solvent [a]	Thiol ligand	Au-NP original size HRTEM	TEM & HRTEM median diameter / nm (standard deviation σ) [a]	dynamic light scattering median diameter / nm (standard deviation σ)[b]
1	Ethanol	HS-CH$_2$-COOH	1.8 (± 0.4)	2.0 (± 0.4)	5.77 (± 1.06) [c]
2	Ethanol	HS-(CH$_2$)$_2$-COOH	1.8 (± 0.4)	---	6.97 (± 1.25) [d]
3	CHCl$_3$	HS-(CH$_2$)$_9$-CH$_3$	1.8 (± 0.4)	4.3 (± 1)	9.53 (± 3.07)
4	CHCl$_3$	HS-(CH$_2$)$_{11}$-CH$_3$	1.8 (± 0.4)	---	10.2 (± 2.81)
5	CHCl$_3$	HS-(CH$_2$)$_{13}$-CH$_3$	1.8 (± 0.4)	---	11.3 (± 2.95)

[a] Statistical evaluation of the total sample pictures. TEM and HRTEM show only the Au core cluster. [b] Hydrodynamic radius, median diameter from the first 3 measurements at 633 nm. The hydrodynamic radius is roughly 2-3 times the size of the Au core cluster due to the ligand and solvent shell. For very small Au-NPs (~1 nm) the size of the hydrodynamic radius can even increase to more than 3 times the Au-NP radius. The resolution of the DLS instrument is 0.6 nm. [c] H-bonded agglomerates appear in the range of about ~35 nm. [d] H-bonded agglomerates appear in the range of about ~ 24 nm.

3.5.3 Au-NP deposition on PTFE (Teflon)

The synthesized Au-NPs can also be deposited onto and stabilized by interaction with a polytetrafluoroethylene (PTFE, Teflon) surface. The *in situ* deposition of gold nanoparticles on PTFE was performed using a 10 × 10 mm^2 PTFE sheet or membrane during the Au-NP synthesis and by post-synthetic photolytic treatment from an Au-NP/BMIm$^+$BF$_4^-$ dispersion (Scheme 3.5.4). A controlled deposition of gold nanoparticles onto PTFE (Teflon) could be a great advantage for different applications, e.g. for the development of electronic, catalytic and sensor devices.[183] We choose PTFE as a support material because of its highly thermal stability. A sheet made of polypropylene or of poly(difluoroethylene) (CHF-CHF)$_n$ were found to melt quickly under our reaction temperature of about 250 °C. The decoration of Au nanoparticles on PTFE was analyzed with an electron probe micro analyzer (EPM), that is by scanning electron microscopy (SEM) and energy dispersive X-ray spectroscopy (EDX) (see Fig. 3.5.8). Au-NP deposition on PTFE was obtained reproducibly during the nanoparticle synthesis (*in situ*) by thermal, photolytic or microwave decomposition/reduction under argon from Au(CO)Cl in the presence of PTFE (Fig. 3.5.8c). In addition, Au-NP deposition could be achieved through UV-irradiation of Au-NP/BMIM$^+$BF$_4^-$ dispersions in the presence of PTFE (Fig. 3.5.8d).

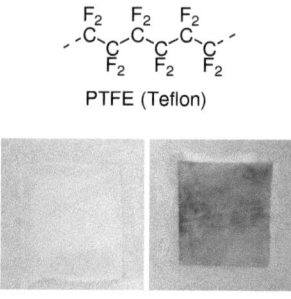

PTFE (Teflon)

Scheme 3.5.4: Surface deposition of Au nanoparticles from BMIm$^+$BF$_4^-$ on a PTFE surface; photograph of the PTFE sheet before (left) and after (right) the Au-NP deposition.

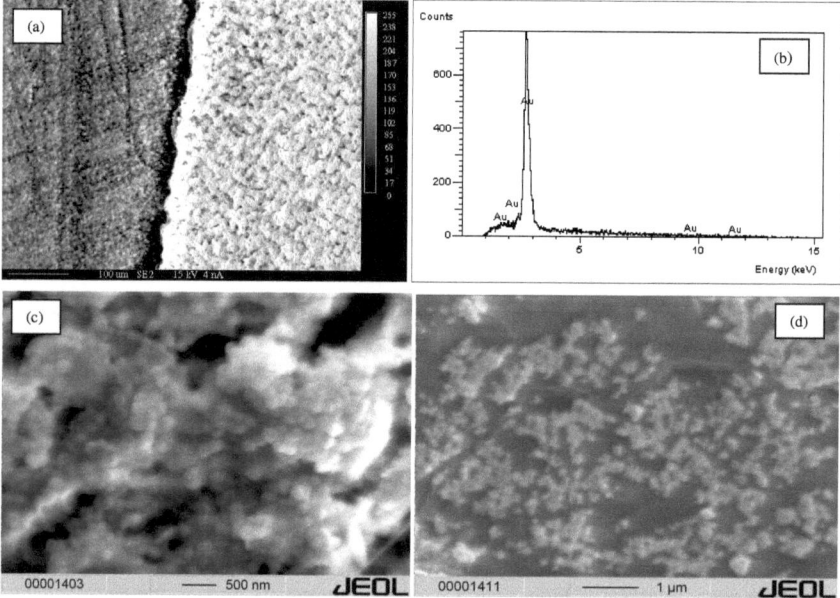

Fig. 3.5.8: (a) EPM-SEM picture of PTFE (Teflon) blank (left part) and its swelling in the presence of BMIm$^+$BF$_4^-$ (right part); b) EDX of Au-NP decorated PTFE surface; c) SEM of Au-NPs on PTFE membrane through *in situ* photolytic deposition from Au(CO)Cl in BMIm$^+$BF$_4^-$. d) SEM of Au-NPs on PTFE sheet through UV-irradiation from Au-NP/BMIm$^+$BF$_4^-$ dispersion.

The PTFE sheet or membrane swells in solvents and also in the ionic liquid BMIm$^+$BF$_4^-$ which leads to a surface change of the PTFE (Fig. 3.5.8a). This swelling contributes to the deposition of the metal nanoparticles. The Au-NPs deposited on the PTFE surface are protected by a thin film of the ionic liquid BMIm$^+$BF$_4^-$ which could also be detected by the EPM measurements.

The best results for the gold decoration of PTFE were obtained by the *in situ* photolytic and microwave assisted decomposition reactions of Au(CO)Cl in BMIm$^+$BF$_4^-$ (Fig. 3.5.8c and 3.5.8d). The *in situ* photolytic decoration of PTFE from Au(CO)Cl/BMIm$^+$BF$_4^-$ results in a high loading of large spherical Au-NPs with a narrow particle size distribution, in comparison to the two other methods, in the range from 170 nm to 230 nm (Fig. 3.5.8c). *In situ* decoration of PTFE from Au(CO)Cl/BMIm$^+$BF$_4^-$ under microwave conditions gives larger Au-NPs with a broader particle size range from 160 nm to 1030 nm. Thermal assisted decoration of PTFE by the decomposition/reduction of Au(CO)Cl results in a lower loading of large particles with a broad particle size distribution in the range from 220 nm to 1020 nm). The photolytic decoration of PTFE with pre-synthesized Au-NPs in BMIm$^+$BF$_4^-$ yields Au-NPs with a particle size distribution in the range from 60 nm to 170 nm (Fig. 3.5.8d). Furthermore, in all gold decorated samples, EDX studies show that the Au nanoparticles consist only of gold without any chlorine (Fig. 3.5.8b).

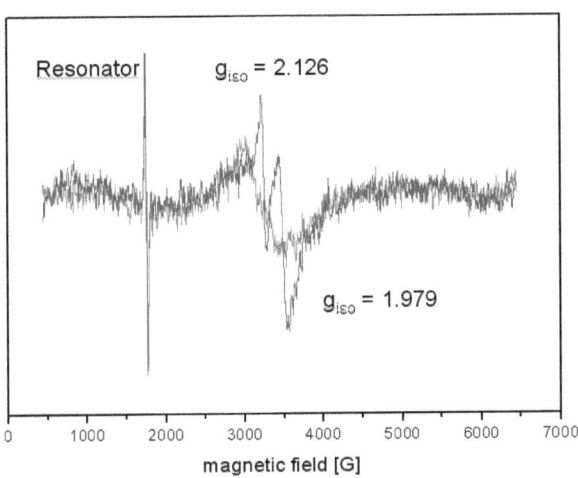

Fig. 3.5.9: ESR signal of the Au decorated PTFE sheet (blue) and blank reference (red).

The Au-NP decorated PTFE shows an EPR resonance with g_{iso} = 2.126 and g_{iso} = 1.979 (Fig. 3.5.9). Bulk Au, Au-NP/IL dispersions, PTFE and the ionic liquid are EPR silent. We suggest that the Au particles assume in part a paramagnetic behavior through a charge transfer to the PTFE surface. Possibly localized holes may be generated in the Au-5d shell from a charge transfer to the PTFE surface, as was shown to S atoms of thiol ligands in forming the Au-S bonds.[202]

3.5.4 Dynamic NMR studies

The interaction of gold nanoparticles with the ionic liquid dynamic matrix was studied by Dr. Harald Scherer by solution nuclear magnetic resonance (^{19}F-NMR / ^{11}B-NMR / ^{1}H-NMR).
The ^{19}F-NMR signal of BMIm$^+$BF$_4^-$ shows a small shift if Au-NPs are dispersed in this solvent. The shift increases with the Au-NP concentration and amounts to 0.02 ppm downfield of the corresponding signal of the pure ionic liquid with about 1 wt.% Au in solution (Fig. 3.5.10).

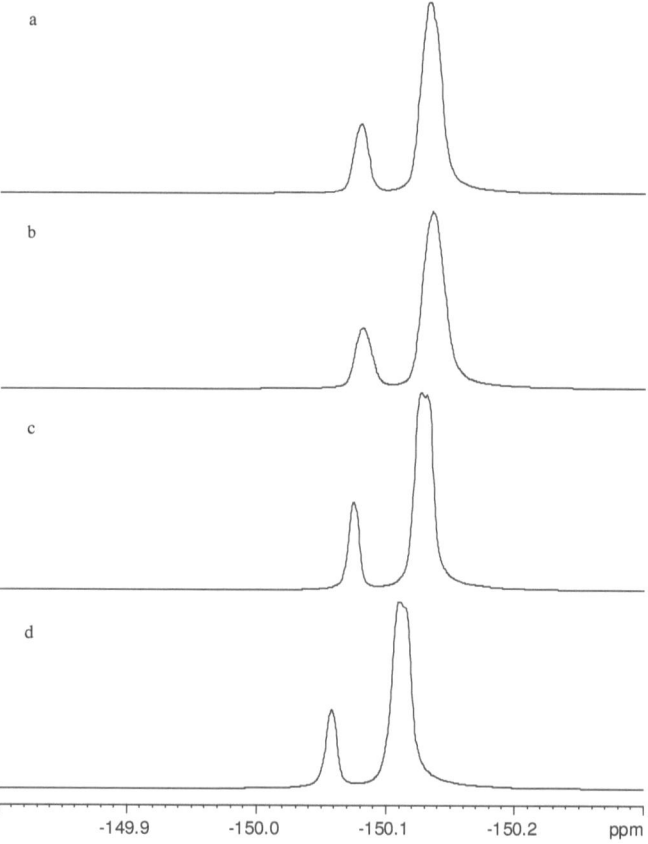

Fig. 3.5.10: ^{19}F-NMR-spectra of BMIm$^+$BF$_4^-$ (without lock, 376.54 MHz) at 25 °C. a: pure ionic liquid, b: BMIm$^+$BF$_4^-$ with 0.25 wt% Au nano particles, c: BMIm$^+$BF$_4^-$ with 0.5 wt% Au nano particles, d: BMIm$^+$BF$_4^-$ with 1.0 wt% Au nano particles.

At the same time the ^{11}B chemical shift of the anion and the ^{1}H chemical shift of the imidazolium cation remain unchanged. Due to the weak Au···F contacts, fluctuation and diffusion will lead to a fast exchange between BF$_4^-$ in the vicinity of Au nanoclusters and the bulk ionic liquid. So, on the NMR timescale there are no different resonances but a small effect on the chemical shift of the average signal. Although the shift difference is quite small for ^{19}F-NMR chemical shifts, the absence of changes of the boron and hydrogen chemical shift makes it unlikely, that the shift difference observed in ^{19}F-NMR is due to the modification of the overall magnetic susceptibility of the sample by the Au nanoclusters in solution. Thus, this finding supports the idea that there are contacts between the fluorine atoms of the BF$_4^-$ anions and the surface of the gold nanoparticles.

These observations contribute to the model which predicts the anions as the primary source of stabilization for electrophilic metal nanoclusters. Our NMR studies also agree with the location of PF$_6^-$ anions from BMIm$^+$PF$_6^-$ on a Pd nanocluster surface by XPS.[203] Moreover, the DFT calculations presented in the next section support the suggested Au-NP···F-BF$_3^-$ interactions, which are crucial for the dynamic IL stabilization of the Au-NPs.

3.5.5 DFT calculations

DFT calculations were done through cooperation with Dr. Michael Walter. In IL dispersions the metal nanoparticles are believed to be surrounded by the anions from the ionic liquid that induce a positive charge on the metal particles surface.[204] The origin of this model is the DLVO theory in analogy of the charge stabilization of colloids in the presence of anions in common solvents.[205] On the other hand, a stabilization due to special end groups of the cations was also proposed.[206] We have investigated the binding of the IL anions and cations, as well as chloride as a synthetic byproduct from Au(CO)Cl or KAuCl$_4$ or a possible IL impurity and free imidazoles via density functional theory (DFT) calculations.

Figure 3.5.11 shows Au$_n^-$ anion binding configurations and the variation of the BE (Binding Energy) with cluster size n. The OTf$^-$ anion binds preferably via an O atom and the flourous ions via the F atoms. The energy difference for bonding between just one, two or three F atoms to the gold-cluster is found to be iso-energetic (energy difference <0.1 eV). This behaviour is reflected in the bond distance between gold clusters and anions depending on cluster size. A similar behavior can be observed for chloride anions which are present as a synthetic byproduct from Au(CO)Cl or KAuCl$_4$ and as a contamination from the IL synthesis process. Remarkably, the chloride anion shows the largest BE of all anions and can, hence, be expected to be bound to the clusters if it is present in the dispersion. In contrast to the case of the anions, the BE of the cations (BMIm$^+$ and dimethylimidazolium, dMIm$^+$) is much lower than the BE of the anions (< 0.5 eV). The free imidazole bases methylimidazole (MIm) and n-butylimidazole (BIm), however, show binding energies similar to the anions (~ 1.5 – 1.7 eV), both in strength as well as in cluster size dependence as shown in Fig. 3.5.11. This supports the model of stabilization enhancement due to free bases proposed recently.[182]

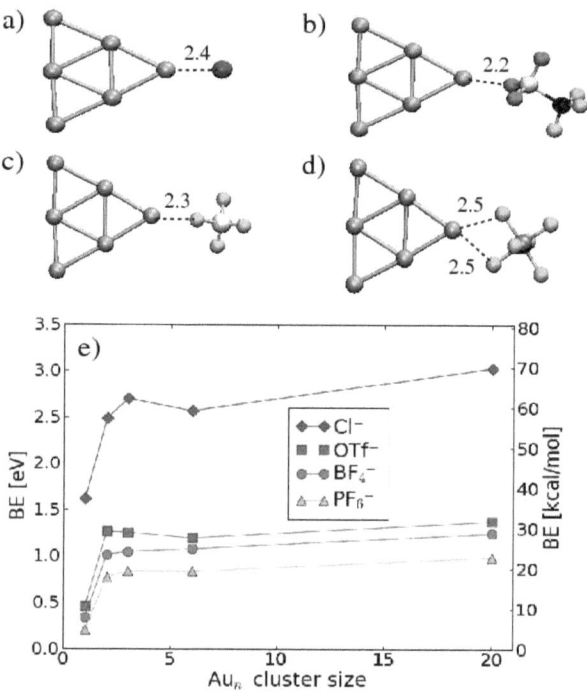

Figure 3.5.11: Relaxed configurations of Au$_6$ bound to a) Cl$^-$, b) OTf$^-$, c) BF$_4^-$ and d) PF$_6^-$. The bond lengths are given in Å. e) Binding energy of different anions depending on the Au$_n$ cluster size.

Chapter 3.6: Stepwise, Ligand-Free and Controlled Growth of Gold Nanoparticles in Ionic Liquids (ILs)

The step by step, ligand-free controlled growth of Au-NPs was prepared by the *in situ* salt-reduction of $KAuCl_4$ with $SnCl_2$ in different ILs.

Ionic liquids (ILs) as a "*nanosynthetic template*" stabilize metal nanoparticles on the basis of their high ionic charge, high polarity, high dielectric constant and supramolecular network.[207,208,209] So that no extra stabilizing molecules or organic solvents are needed.[210] Recently, fast and ligand stabilized Au particle growth in common organic solvents was monitored by synchrotron SAXS/WAXS[211] or optical single particle spectroscopy[212] for mechanism and kinetic studies of the Au nanoparticle nucleation and growth process.[213]

The Au growth process can be controlled through the drop wise addition of a $KAuCl_4$ solution to a mixture of $SnCl_2$/IL. (Eqn. 1)

$$2\,[KAuCl_4] + 3\,SnCl_2\ \underset{RT}{\overset{ILs}{\longrightarrow}}\ 2\,Au(NPs) + 2\,SnCl_4 + K_2[SnCl_6] \quad (1)$$
$$\text{(excess)}$$

During the addition of $KAuCl_4$/IL to excess $SnCl_2$/IL the color changes from light-yellow, to yellow and yellow-orange, orange-red, red-purple, purple-brown, brown-black and finally black-blue (Figure 3.6.1-3.6.5). The color correlates directly with the Au-NP size (Fig. 3.6.4).[214] The Au-NP growth process can reproducibly be controlled and stopped at any color step, this is Au particle size, by discontinuing addition of $KAuCl_4$.

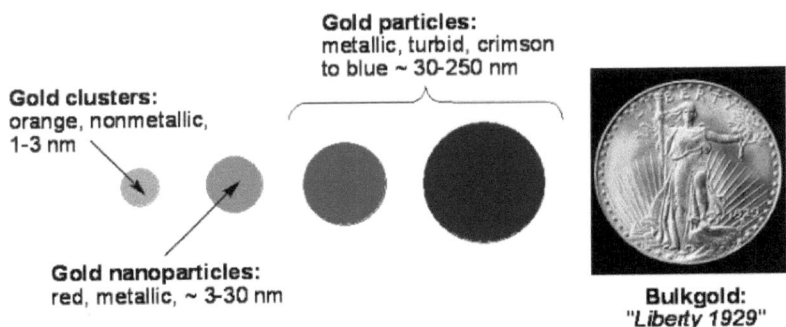

Fig. 3.6.1: Colour change and correlation of Au-NPs with the size.

Monitoring the Au-NP growth by TEM/HRTEM and UV/VIS spectroscopy (Figure 3.6.2. and 3.6.5) in different ionic liquids shows a significant peak shift in the UV/VIS spectrum to longer wave lengths (>520 nm) during addition of KAuCl$_4$, which we ascribe to the formation of larger (>4 nm) and simultaneously crystalline and metallic Au-NPs.[215] Still, the short wave-length maximum at 360-380 nm, assigned to 1-3 nm amorphous and non crystalline Au-NPs[215] is retained, so that UV/VIS shows not only the growth to larger particles, but also the formation of new small Au-NPs during the addition of KAuCl$_4$. This is supported by HRTEM and TEM measurements (Table 3.6.1) where a bimodal particle size distribution, due to the formation of new small particles, develops upon continuous KAuCl$_4$ addition when the median larger particle size surpasses ~ 5 nm.

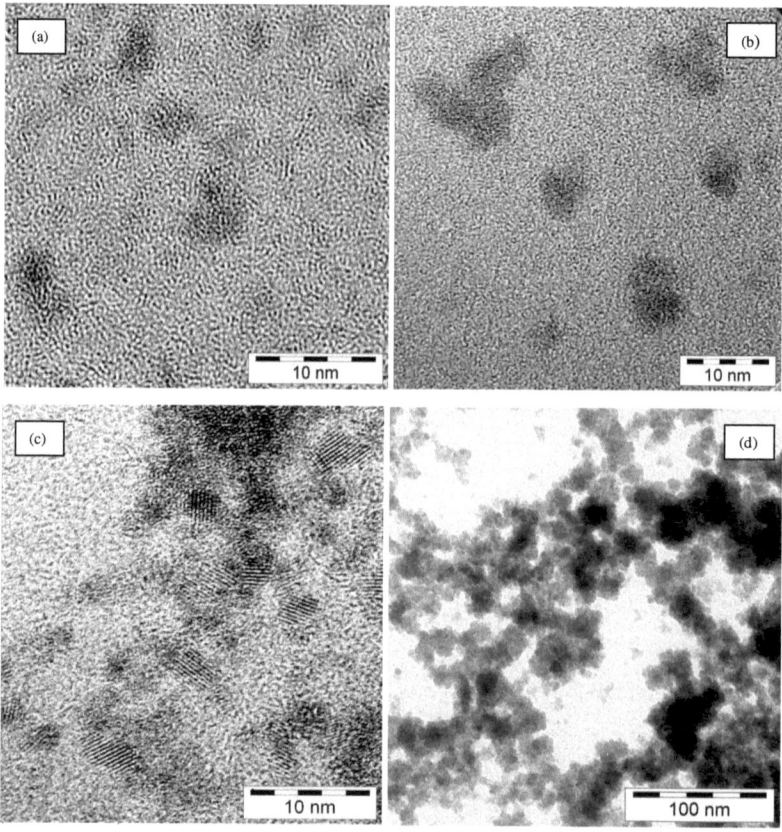

Fig. 3.6.2: HRTEM and TEM pictures of Au-NPs from KAuCl$_4$/SnCl$_2$ in BMIm$^+$BF$_4^-$; (a) no. 1; (b) no. 2; (c) no. 3; (d) no. 6 in Table 3.6.1; crystalline Au-NPs show lattice layers, see TED in Fig. 3.6.3.

From transmission electron diffraction (TED) und absorbance UV/VIS studies Fig. 3.6.3 and Fig. 3.6.5, we observed a critical transition point from 1-3 nm amorphous and nonmetallic Au-NPs to crystalline and metallic Au-NPs (~ 3-5 nm) at a molar Au:Sn ratio of about 1:10 (see Fig 3.6.5), which is indicated by an increase in the gold plasmon band at around 520 nm. The median metal nanoparticle size depends on the added Au content, which ranges between ~ 1.7 to max. 60 nm. (see, Fig. 3.6.4 and Table 3.6.1).

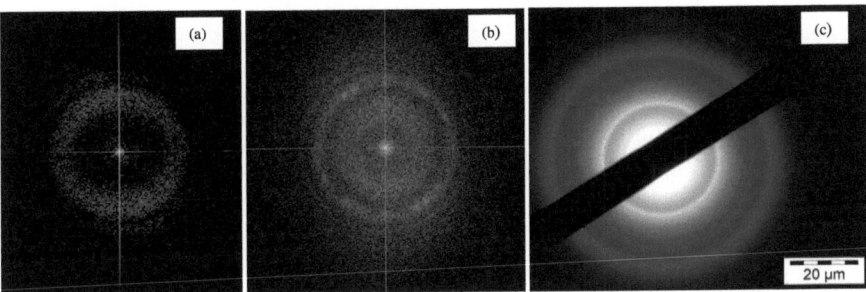

Fig. 3.6.3: TEDs from HRTEM and TEM from Fig. 3.6.2, in comparison (a) no. 2; (b) no. 3 and (c) no. 5 from Table 3.6.1. The diffraction rings at (Å) 2.4 (very strong), 2.1 (strong), 1.5 (medium), 1.3 and 1.0 (all weak) match with D spacing of the Au diffraction pattern.[173]

Table 3.6.1: Au-NP size, distribution and range in BMIm$^+$BF$_4^-$ analyzed by (HR)TEM during the controlled growth.

No.	V(KAuCl$_4$/IL) / molar Au:Sn ratio	TEM / HRTEM median diameter / nm (standard deviation σ)[a,b]		TEM / HRTEM size range / nm	
1	0.15 mL / 1:24	2.6 ± 0.6		1.7 – 4.0	
2	0.3 mL / 1:12	4.1 ± 1.0		2.3 – 5.6	
3	0.6 mL / 1:6	4.9 ± 1.4	1.9 ± 0.4	4.4 – 6.8	1.3 – 2.6
4	0.7 mL / 1:5	6.5 ± 1.5	3.2 ± 0.6	4.4 – 11	2.1 – 4.9
5	0.8 mL / 1:4.4	7.6 ± 2.3	3.6 ± 0.7	5.0 – 16	2.0 – 5.5
6	0.9 mL / 1:4	22 ± 6	4.1 ± 0.4	6.9 – 32	2.0 – 8.5
7	1.2 mL / 1:3	36 ± 10	5.9 ± 1.3	20 – 60	2.7 – 9.3

30.0 mg (0.08 mmol) KAuCl$_4$ in 3.0 g (2.5 ml) IL, c(Au) = 0.03 mmol/mL, from which a specified amount was added to 20.0 mg (0.106 mmol) SnCl$_2$ in 2.0 g (1.7 ml) IL, BMIm$^+$BF$_4^-$ IL density 1.208 g/cm^3. [a] Statistical evaluation of the total sample pictures (see Supporting Information). (HR)TEM = (high resolution) transmission electron microscopy. [b] bimodal particle distribution for No. 3-7.

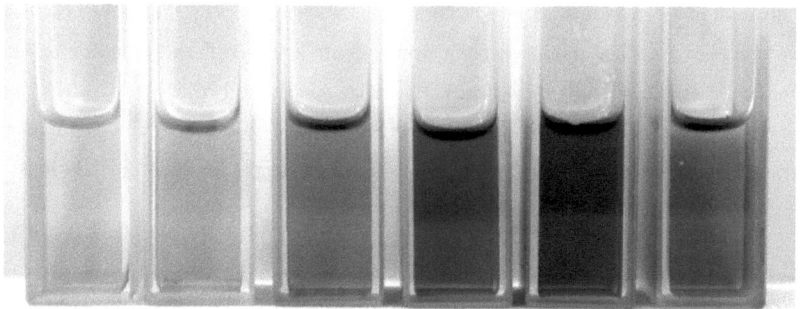

Fig. 3.6.4: Selected colors during the Au growth process, controlled by the (stated) molar Au:Sn ratios (from left to right) 1:24; 1:12; 1:10; 1:8; 1:6; 1:4.5. (See, also UV/VIS spectra Fig. 3.6.5) Transition of Au-NPs from non-metallic (yellow) particles to metallic and crystalline (red and purple) Au-NPs.

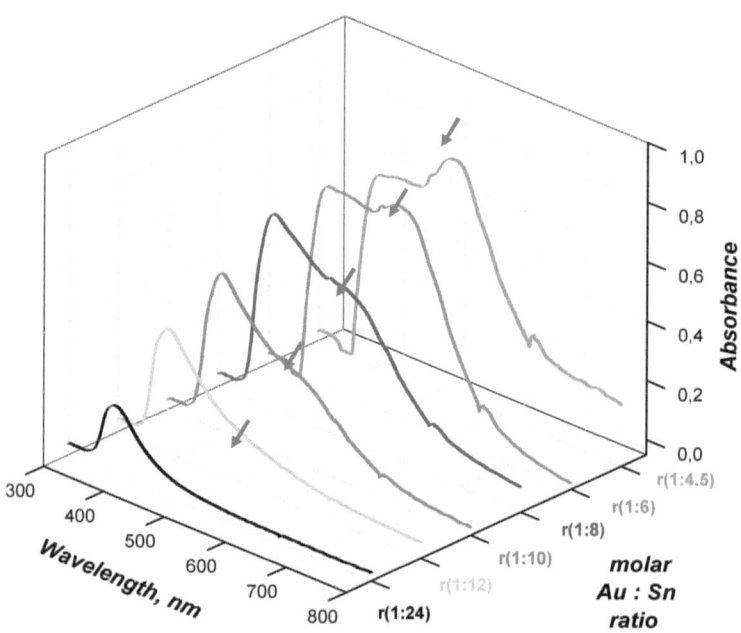

Fig. 3.6.5: 3D UV/VIS spectra of Au-NPs from $KAuCl_4/SnCl_2$ in $BMIm^+BF_4^-$. Selected colors during the growth process, controlled by the (stated) molar Au:Sn ratios 1:24; 1:12; 1:10; 1:8; 1:6; 1:4.5, (Fig. 3.6.4) Arrows (green) indicate the shift to longer wavelengths by the increase of the Au plasmon band.

Nucleation and Growth Mechanism of Au-NPs

Beginning with freshly formed Au metal atoms nucleation occurs to a gold seed/nuclei (beginning with linear Au_{1-3} to planer Au_{4-10} and ending with 3D Au_{11-20} species)[216]. Adding more Au metal atoms to the Au seeds/nuclei growth occurs via deposition of Au atoms onto the solid surface (molecular addition). Moreover, there is a different growth rate for bigger and smaller Au-NPs at this stage where smaller particles will grow faster on account of their bigger free energy. After reaching a critical concentration secondary growth can occur by aggregation with other Au particles. However, the growth by this process is faster than by molecular addition, and it occurs by stable particles combining with smaller unstable seeds/nuclei. Finally, the growth process is commonly stopped by the presence of a surface-capping ligand.[210] Thus, ionic liquids are not acting as strong capping ligands, the Au growth can be controlled in an easy and efficient way. (Fig. 3.6.4 and Fig. 3.6.5)

This Au-NP growth process can be stopped at every intermediate color step, which allows a careful investigation of the Au nuclei's and nanocrystal formation.

Chapter 4: Conclusions

This thesis describes the reproducible synthesis of different transition Metal Nanoarticles (M-NPs), with M = Cr, Mo, W, Fe, Ru, Os, Co, Rh, Ir, Ag and Au, from different precursors (metal salts, metal complexes and organometallic compounds) using different synthetic methods (H_2 reduction, in situ salt reduction, thermolytic, photolytic and microwave based decomposition and reduction processes) in Ionic Liquids (ILs).

The intrinsic properties of ILs, such as their ionic charge, high polarity, high dielectric constant, high thermal stability and low reactivity were seen as advantageous for their use as a nanosynthetic medium or template to prepare M-NPs. The ILs act as non-surfactant, or weakly coordinating supramolecular network for the kinetic stabilization of the M-NPs. The size of different transitions M-NPs was shown to depend upon the choice of the IL (especially the anion therein), which was demonstrated using different methods and synthesis procedures.

Stable Ag-NPs were reproducibly obtained, size dependent by the choice of the IL-anion, by H_2 reduction of different Ag(I)X salts (X = BF_4, PF_6, OTf) dissolved ILs in the presence of *n*-butyl-imidazole (Bim) as a scavenger for the HX acid byproduct in order to avoid disturbance of the IL network and, thus Ag-NP destabilization (Chapter 3.1).

Furthermore, Cr-, Mo- and W-NPs were obtained size-dependent, through the choice of the IL-Anion, by the thermal and photolytic decomposition of the corresponding metal carbonyls $M(CO)_6$ (M = Cr, Mo, W) (Chapter 3.2).

Metal carbonyls were introduced as precursors for M-NP synthesis in ILs to avoid problems like the forming of acid unwanted byproducts during H_2 reduction of metal salts. In addition stable Fe-, Ru-, Os-, Co-, Rh- and Ir-NPs were obtained reproducibly by thermal or photolytic decomposition under argon from the metal carbonyl precursors $Fe_2(CO)_9$, $Ru_3(CO)_{12}$ or $Os_3(CO)_{12}$, $Co_2(CO)_8$, $Rh_6(CO)_{16}$ and $Ir_4(CO)_{12}$, suspended in the ILs $BMIm^+BF_4^-$, $BMIm^+OTf^-$ and $BtMA^+NTf_2^-$ respectively (Chapters 3.3 and 3.4).

The very small and uniform nanoparticle size of about 1 to 3 nm in $BMIm^+BF_4^-$ increased with the molecular volume of the IL anion in $BMIm^+PF_6^-$, $BMIm^+OTf^-$ and $BtMA^+NTf_2^-$. Characterization of the dispersed transitions nanoparticles was done by transmission electron microscopy (TEM and HRTEM, size determination), transmission electron diffraction (TED, crystallinity and verification of metal state versus metal oxide identity), X-ray Powder Diffraction (XRPD, crystallinity, metal state identity and size approximation from the Scherrer equation) and Dynamic Light Scattering (DLS, size determination and hydrodynamic radius). Under argon the M-NP/IL dispersions were kinetically stable without any additional stabilizers or capping molecules.

Gold nanoparticles (Au-NPs) were prepared by thermal decomposition and reduction of Au(CO)Cl or $KAuCl_4$ in the presence of *n*-butyl-imidazole as scavenger in different ILs. The Au-NPs can then be transferred to polar and non-polar organic solvents by introducing organic capping molecules like thiolglycolic acid and n-decanethiol or onto a polytetrafluoroethylene (PTFE) surface (Chapter 3.5).

Furthermore, Nuclear Magnetic Resonance (NMR) spectroscopy and Density Functional Theory (DFT) calculations were done with different corporation partners to understand the interactions and stabilization mechanisms of dispersed transition M-NPs in ILs. Binding and stabilization of Au-NPs with the IL anions was confirmed by DFT calculations, and was found to be much stronger than binding to cations. The DFT calculations and dynamic NMR studies support strongly IL/anion···Au-NPs interactions and a dynamic anion stabilization around M-NP clusters, which is caused by the low binding interaction of weakly coordinating IL-Anions (Chapter 3.5). This mechanistic understanding was crucial for a deep insight into the interaction processes between the IL-network and the "naked" metal nanoclusters.

The sufficient, but not too strong, stabilizing network properties of ILs towards M-NPs allowed us to succeed in step-by-step sequential growth of Au-NPs in ILs, which was monitored by the specific colour changes and an increase in the Au-plasmon band for metallic Au nanoparticles during nucleation and growth. (Chapter 3.6)

This work can be described as a novel, facile and general method for preparing of transition M-NPs, as distinct from the common colloidal synthesis routes. This work sets the stage for further research in this field. Two diploma students Jérome Krämer and Christian Vollmer continue research in this area. Furthermore, two DE-Patent applications have now been officially accepted. Based on the DE-Patent applications two international PCT/EP/WO-Patent applications have been further submitted.

Chapter 5: Instruments & Experimental Section

5.1 Instrumentation and Devices

Transmission electron microscopy (TEM and TED) and high resolution transmission electron microscope (HRTEM) micrographs were taken at room temperature from a carbon-coated copper grid, at the FMF (Univ. Freiburg) by Dr. R. Thomann using of a Zeiss LEO 912 (TEM) transmission electron microscope operating at an accelerating voltage of 120 kV. HRTEM micrographs were taken at the Laboratory of Electron Microscopy (LEO) at Univ. of Karlsruhe (TH) by Dr. habil. Schneider on a HRTEM Phillips CEM 200 ST operating at an accelerating voltage of 200 kV. (Fig. 5.1)

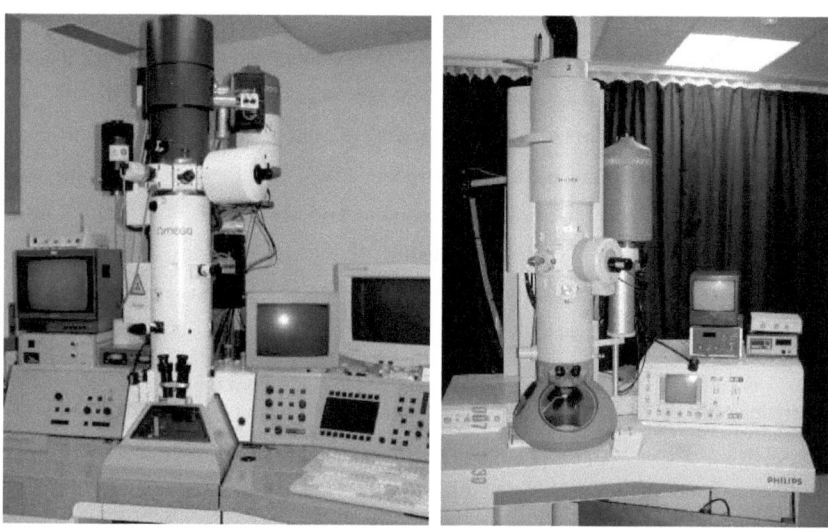

Fig. 5.1: TEM Zeiss LEO 912 (left) and HRTEM Phillips CEM 200 ST (right)

A Malvern Zetasizer Nano-ZS was used for the dynamic light scattering measurements working at 633 nm wavelength (Fig. 5.2a). Care was taken for choosing the right parameters, such as the index of refraction of the corresponding transition metal at 0.1 absorption at this wavelength. Samples were prepared by dilution of 10 µL (1wt. % of M) of the metal/IL dispersion in 2 mL n-butyl-imidazole (99% p.a.; particle free). 150 µL of this M/IL/n-butyl-imidazole solution were diluted further with 1.5 mL of n-butyl-imidazole and use it in a glass cuvette before measurement.

UV/VIS absorption measurement on Au-NP dispersions were done with a J&M *TIDAS* UV/VIS spectrometer in the wavelength range between 200 and 800 nm (Fig. 5.2b).

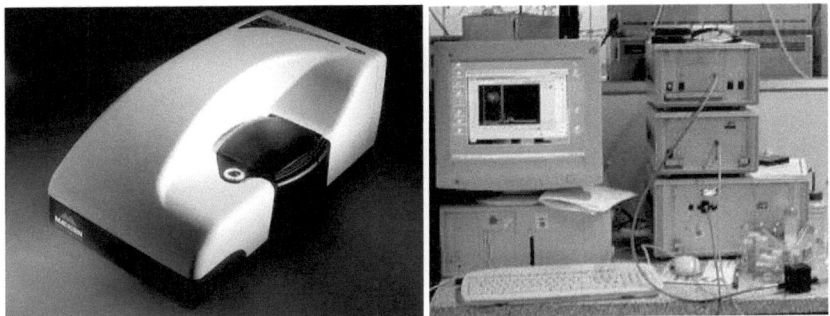

Fig. 5.2: a) Picture of a Malvern Zetasizer Nano-ZS; b) TIDAS UV/VIS spectrometer.

Scanning electron microscopy (SEM) photographs were taken at room temperature from a *JOEL* JSM-6300F scanning microscope operating at an accelerating voltage of 5 kV.

Energy dispersive spectrometry (EDS) measurements were performed on a CAMECA SX100, electron probe micro analyzer (EPMA) equipped with an Oxford Penta FET detector, at room temperature with an accelerating voltage of 15 kV.

NMR experiments were carried out on a BRUKER AVANCE II WB spectrometer with a 5 mm ATM-BBFO probe head at room temperature. The spectrometer was locked and shimmed on a sample containing a mixture of $BMIm^+BF_4^-$ and $CDCl_3$. The subsequent samples were detected without lock. To avoid field shifts by shim gradients, these samples were not further shimmed. The field drift of the magnet was 0.5 Hz per hour.

Electron spin resonances (ESR) measurements were measured on an ESP 300E at room temperature in glass tube under continuous wave (CW) conditions.

Powder X-ray diffractograms were measured at ambient temperature using a STOE STADI-P with Debye-Scherrer geometry, Mo-Kα radiation (λ = 0.7093 Å), a Ge(111) monochromator and the samples in glass capillaries on a rotating probe head. Simulated powder patterns were based on single-crystal data and calculated using the STOE WinXPOW software package.

5.2 Synthesis equipment

The H_2 salt reductions were carried out at 4 atm (constant pressure) in a stainless steel reactor (Büchi-Miniclave) connected to an H_2 gas cylinder. (Fig. 5.3)

Fig. 5.3: Büchi-Miniclave reactor for H_2 salt reductions.

Photolytic decompositions of a dispersion of $M_x(CO)_y$/IL were carried out in a Kürner UV 1000 reactor from Kürner-Analysentechnik in quarz tubes for 1-15 min under argon with a Hg-UV lamp (1000 W) in the range 200-450 nm. (Fig. 5.4)

Fig. 5.4: Kürner UV 1000 reactor for photolytic decompositions.

Thermal decompositions were carried out under argon (or air) in a vessel which was connected to an oil bubbler. The dispersions $M_x(CO)_y$/IL was heated up to 180-250 °C for several hours (3.5-18 h) to thermally decompose the corresponding metal carbonyl. (Fig. 5.5)

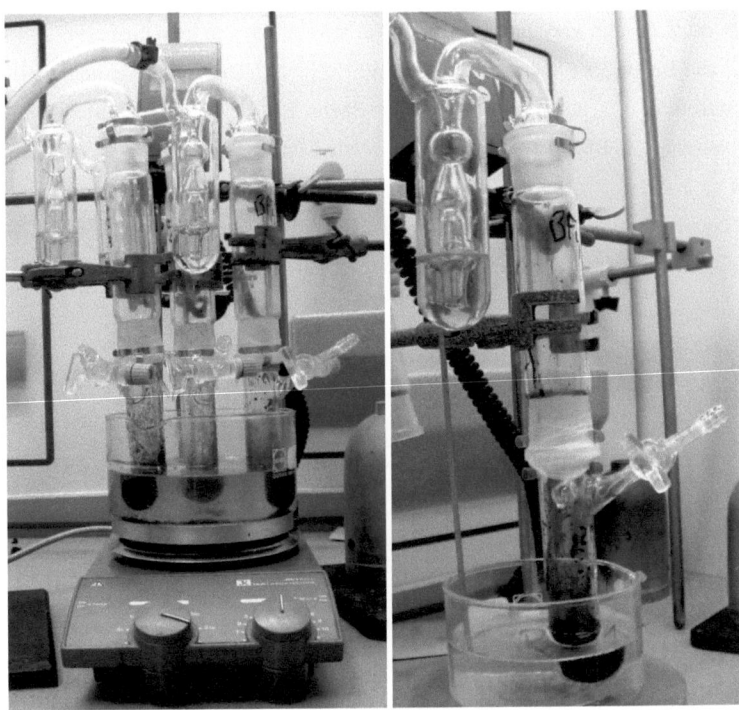

Fig. 5.5: Equipment for simple thermal decomposition of metal carbonyls in ionic liquids (ILs).

5.3 Synthesis procedure and details

All manipulations were done using Schlenk techniques under argon since the metal carbonyls and salts are hygroscopic and air sensitive. Some manipulations were done in the glove box. The ionic liquids were dried at high vacuum (10^{-3} mbar) for several days to avoid especially in the case of BMIm$^+$BF$_4^-$ and BMIm$^+$PF$_6^-$ the hydrolysis to HF.[217,218,219,220] The used ionic liquids (ILs) n-Butyl-Methyl-Imidazolium tetrafluoroborate BMIm$^+$BF$_4^-$, n-Butyl-Methyl-Imida-zolium hexafluorophospahte BMIm$^+$PF$_6^-$, n-Butyl-Methyl-Imidazolium trifluoromethane-sulfonate BMIm$^+$OTf$^-$ and n-Butyl-tri-Methyl-Ammonium N-bis(trifluoromethyl-sulfonyl)imide BtMA$^+$Tf$_2$N$^-$ and n-Butyl-Imidazole BIm were delivered from IoLiTec with a certified H$_2$O and Cl$^-$ content (H$_2$O content << 100 ppm; Cl$^-$ content << 50 ppm).

Synthesis of Ag-NPs (Chapter 3.1)

Materials: Ag(I)X salts (X = BF$_4$, PF$_6$, OTf) were obtained from Aldrich and Merck.

The H$_2$ reductions were carried out in a stainless steel reactor connected to an H$_2$ gas cylinder (Fig. 5.3). In a typical experiment the Ag(I)salt (0.0142-0.0426 g) and BIm (0.02-0.05 g) were dissolved and dispersed (~15 min) under argon at room temperature in 3 g of the ionic liquid (0.3-1 wt.% Ag). Then, the reactor was evacuated and placed in an oil bath at 85 °C under stirring. The reaction was started by pressurizing with H$_2$ to 4 atm. After 2 h the reactor was evacuated for 1 h at 100 °C to remove excess H$_2$ and BIm. After cooling to room temperature an aliquot of the ionic liquid was collected under argon atmosphere for *in situ* TEM characterization. The neutral pH of the reaction medium was ascertained by testing with pH paper.

Ag precursor	Ionic Liquid	wt.% Metall	weight (mg)	reduction 4 atm H$_2$
AgBF$_4$	[BMIm]$^+$[BF$_4$]$^-$	0.5	54	75 °C
AgPF$_6$	[BMIm]$^+$[PF$_6$]$^-$	0.3	24	75 °C
AgOTf	[BMIm]$^+$[OTf]$^-$	0.5	36	75 °C
Ag$_2$O	[BMIm]$^+$[BF$_4$]$^-$	0.5	16	75 °C
Ag$_2$O	[BMIm]$^+$[BF$_4$]$^-$	1	32	75 °C
AgBF$_4$	[BMIm]$^+$[BF$_4$]$^-$	0.5	54	75 °C
AgPF$_6$	[BMIm]$^+$[PF$_6$]$^-$	0.5	35	75 °C
AgOTf	[BMIm]$^+$[OTf]$^-$	1	72	75 °C
AgOTf	[BtMA]$^+$[NTf$_2$]$^-$	0.5	36	75 °C
Ag$_2$O	[BMIm]$^+$[BF$_4$]$^-$	0.5	16	75 °C

Synthesis of Cr, Mo and W-NPs (Chapter 3.2)

Materials: $M(CO)_6$ (M = Cr, Mo, W) were obtained from Aldrich and Merck.

In a typical experiment the metal carbonyl $M(CO)_6$ was dissolved/suspended under argon/air in the dried and deoxygenated ionic liquid. The solution was heated up to 230 °C for several hours to thermally decompose the metal carbonyl.

Thermal decompositions were carried out under argon (or air) in a vessel which was connected to an oil bubbler (Fig. 5.5). In a typical experiment the metal carbonyl $M(CO)_6$ (M = Cr, Mo and W, 0.118, 0.078 and 0.057 g, respectively) was dissolved (~1 h) under argon at room temperature in 3.0 g of the ionic liquid to give a 1 wt% M solution. The solution was slowly heated up to 230 °C for 12 h under stirring. After cooling to room temperature under argon an aliquot of the ionic liquid was collected under argon atmosphere for *in situ* TEM and dynamic light scattering (DLS) characterization.

Alternatively, the suspension of $Mo(CO)_6$ (0.008 g) in 3.0 g $BMIm^+BF_4^-$ was irradiated at 200-450 nm for 15 min under argon for photolytic decomposition of $M(CO)_6$ in a Kürner UV 1000 reactor from Kürner-Analysentechnik (Fig. 5.4).

Carbonyl	Ionic Liquid	wt.% Metall	weight (mg)	decomposition
$W(CO)_6$	$[BMIm]^+[BF_4]^-$	1	57	230 °C
$W(CO)_6$	$[BMIm]^+[OTf]^-$	1	57	230 °C
$W(CO)_6$	$[BMIm]^+[NTf_2]^-$	1	57	230 °C
$W(CO)_6$	$[BtMA]^+[NTf_2]^-$	1	57	230 °C
$W(CO)_6$/air	$[BtMA]^+[NTf_2]^-$	1	57	230 °C
$Mo(CO)_6$	$[BMIm]^+[BF_4]^-$	1	78	230 °C
$Mo(CO)_6$	$[BMIm]^+[BF_4]^-$	0.1	8	hv (15 min)
$Mo(CO)_6$	$[BtMA]^+[NTf_2]^-$	1	78	230 °C
$Mo(CO)_6$ / air	$[BtMA]^+[NTf_2]^-$	1	78	230 °C
$Cr(CO)_6$	$[BMIm]^+[BF_4]^-$	1	118	230 °C
$Cr(CO)_6$	$[BtMA]^+[NTf_2]^-$	1	118	230 °C
$Cr(CO)_6$ / air	$[BtMA]^+[NTf_2]^-$	1	118	230 °C

Synthesis of Fe, Ru and Os-NPs (Chapter 3.3)

Materials: $Fe_2(CO)_9$, $Ru_3(CO)_{12}$, and $Os_3(CO)_{12}$ were obtained from Strem.

Thermal decompositions were carried out under argon (air) in a vessel which was connected to an oil bubbler (Fig. 5.5). In a typical experiment the metal carbonyl $M_n(CO)_m$ (M = Fe, Ru and Os, 0.020, 0.014 and 0.008 g, respectively) was dissolved/suspended (~ 1 h) under argon at room temperature in 3.0 g of the ionic liquid to give a 0.2 to 1 wt.% M solution. The solution was slowly heated to 180 – 200 °C for Fe and Ru and to 250 °C for Os in a period of 12 h under stirring. After cooling to room temperature under argon an aliquot of the ionic liquid was collected under argon atmosphere for *in situ* TEM and dynamic light scattering (DLS) characterization.

Photolytic decompositions of a suspension of $Ru_3(CO)_{12}$ (0.005 g) in 3.0 g $BMIm^+BF_4^-$ were carried out in a Kürner UV 1000 reactor from Kürner-Analysentechnik in quarz tubes for 15 min under argon with a Hg-UV lamp (1000 W) in the range 200-450 nm (Fig. 5.4).

Carbonyl	Ionic Liquid	wt.% Metall	weight (mg)	decomposition
$Fe_2(CO)_9$	$[BMIm]^+[BF_4]^-$	0.2	19	180 °C
$Fe_2(CO)_9$	$[BMIm]^+[BF_4]^-$	1	97	180 °C
$Fe_2(CO)_9$	$[BMIm]^+[BF_4]^-$	0.2	19	180 °C
$Fe_2(CO)_9$	$[BMIm]^+[BF_4]^-$	1	97	180 °C
$Ru_3(CO)_{12}$	$[BMIm]^+[BF_4]^-$	0.2	13	200 °C
$Ru_3(CO)_{12}$	$[BMIm]^+[BF_4]^-$	0.6	38	200 °C
$Ru_3(CO)_{12}$	$[BMIm]^+[BF_4]^-$	1	64	200 °C
$Ru_3(CO)_{12}$	$[BMIm]^+[BF_4]^-$	0.08	5	hv (15 min)
$Os_3(CO)_{12}$	$[BMIm]^+[BF_4]^-$	0.2	9,4	250 °C
$Os_3(CO)_{12}$	$[BMIm]^+[BF_4]^-$	1	48	250 °C

Synthesis of Co, Rh and Ir-NPs (Chapter 3.4)

Materials: $Co_2(CO)_8$, $Rh_6(CO)_{16}$, and $Ir_4(CO)_{12}$ were obtained from Strem.

Thermal decompositions were carried out under argon in a glass vessel which was connected to an oil bubbler (Fig. 5.5). In a typical experiment the metal carbonyl was dissolved/dispersed (during ~1 h) under argon at room temperature in 3.0 g of the ionic liquid to give a solution/suspension with a defined weight percentage (wt%) in metal. The solution was slowly heated to 180 °C (Co) or 230 °C (Rh, Ir) for 18 h under stirring. After cooling to room temperature under argon an aliquot of the ionic liquid was collected under argon atmosphere for *in situ* TEM, TED and dynamic light scattering characterization.

Carbonyl	Ionic Liquid	wt.% Metall	weight (mg)	decomposition
$Co_2(CO)_8$	$[BMIm]^+[BF_4]^-$	0.2	14	180 °C
$Rh_6(CO)_{16}$	$[BMIm]^+[BF_4]^-$	0.2	11	230 °C
$Rh_6(CO)_{16}$	$[BMIm]^+[BF_4]^-$	1	52	230 °C
$Rh_6(CO)_{16}$	$[BMIm]^+[BF_4]^-$	1	52	230 °C
$Ir_4(CO)_{12}$	$[BMIm]^+[BF_4]^-$	0.2	9	230 °C
$Ir_4(CO)_{12}$	$[BMIm]^+[BF_4]^-$	0.2	9	230 °C
$Ir_4(CO)_{12}$	$[BMIm]^+[BF_4]^-$	1	43	230 °C
$Ir_4(CO)_{12}$	$[BMIm]^+[OTf]^-$	0.5	21	230 °C
$Ir_4(CO)_{12}$	$[BtMA]^+[NTf_2]^-$	0.5	9	230 °C

Synthesis of Au-NPs (Chapter 3.5)

Materials: Au(CO)Cl and KAuCl$_4$ were obtained from STREM. PTFE membranes cut to 10 × 10 mm^2 per piece from the 45 mm diameter commercially obtained membranes from Sartorius Stedim Biotech GmbH, 65 μm thickness, pore size 0.45 μm; PTFE pieces cut to 10 × 10 mm^2 sections from PTFE sleeves intended to NS 29 glass joints, 0.05 mm thickness from VWR International GmbH.

Gold nanoparticle syntheses in ILs

a) By thermal decomposition: Thermal decompositions were carried out under argon in a glass vessel which was connected to an oil bubbler (Fig. 5.5). In a typical experiment 0.040 g (0.154 mmol) of Au(CO)Cl was dissolved/dispersed (during ~1 h) under argon at room temperature in the presence of 1.5 equivalents (0.231 mmol) of *n*-butyl-imidazole in 3.0 g of the ionic liquid. The solution was slowly heated to 230 °C for 18 h under mechanical stirring. The decomposition process gave a dispersion with 0.03 g gold metal (1 wt.% Au-NP in IL). During the decomposition process a white haze of *n*-butyl-imidazolium chloride is formed. After cooling to room temperature under argon a white-yellow precipitate (of *n*-butyl-imidazolium chloride, see below) was obtained and an aliquot of the ionic liquid was collected under argon atmosphere for *in situ* TEM/HRTEM, TED and dynamic light scattering (DLS) characterization. The white-yellow precipitate was collected by centrifugation (13200 rpm for 5 min) with decanting the supernatant Au/IL dispersion. The precipitate was washed twice with CHCl$_3$ (2 mL) and dried for several days at high vacuum. The dried white-yellow precipitate could be characterized as *n*-butyl-imidazolium chloride by elementary analysis: Found C 52.69, H 8.28, N 17.56; calculated C 52.34, H 8.16, N 17.44%.

^{1}H-NMR (200 MHz, (CD$_3$)$_2$CO, 20 °C): δ/ppm (TMS as internal standard) 9.02 (br, 1H, Aryl-N-CH-N), 8.6 (vbr, 1H, Aryl-NH), 7.67 (t, J = 1.5, 1H, Aryl-N-CH), 7.49 (t, J = 1.5, 1H, Aryl-N-CH), 4.4 (t, J = 7.3, 2H, NCH$_2$), 1.97 (m, J = 7.4, 2H, CH$_2$), 1,41 (m, J = 7.4, 2H, CH$_2$), 0.98 (t, J = 7.4, 3H, CH$_3$) (br = broad, s = singlet, d = doublet, t = triplet, m = multiplet, v = very.

b) By photolysis: Photolytic decompositions of a solution/suspension of Au(CO)Cl (0.030 g, 0.115 mmol) in 2.0 g BMIm$^+$BF$_4^-$ were carried out in a Kürner UV 1000 reactor from Kürner-Analysentechnik in quarz tubes for 1 min under argon using a Hg-UV lamp (1000 W) in the range between 200-450 nm to give a dispersion with 0.020 g gold metal (~1 wt.% Au-NP in IL) (Fig. 5.4).

c) By microwave radiation: Microwave decomposition of a solution/suspension of Au(CO)Cl (0.030 g, 0.115 mmol) in 2.0 g BMIm$^+$BF$_4^-$ were carried out in CEM Discovery Microwave in a glass tube for 3 min at 10 Watt (250 °C) under argon without stirring to give a dispersion with 0.020 g gold metal (~1 wt.% Au-NP in IL).

Au-NP surface functionalization

Post synthetic surface functionalization of the dispersed gold nanoparticles in ILs was done by treating the 1wt.% Au-NP/IL dispersion with an excess of different organic thiol ligands. The volume ratio of the neat thiol liquid to the IL liquid was 5:1. The thiolated Au nanoparticles were collected under centrifugation (13200 rpm for 5 min) and then washed twice with 20% (v/v) water-methanol solution (2 mL) in an ultrasonic redispersion-centrifugation process. This process was repeated with 2 mL methanol (p.a.) to remove unbound thiol ligand.

Au-NP deposition on PTFE

It is also possible to carry out the above gold nanoparticle syntheses in ILs in the presence of a piece of PTFE for *in situ* Au-NP deposition onto PTFE. By adding a piece of PTFE to the thermal decomposition, the photolysis or microwave decomposition process the *in situ* gold deposition onto PTFE was obtained as a purple-red thin film.

Post synthetic photolytic decoration of a 10×10 mm^2 PTFE piece with pre-synthesized Au-NPs from BMIm$^+$BF$_4^-$ were carried out in a Kürner UV 1000 reactor from Kürner-Analysentechnik in quarz tubes for 1 min under argon using a Hg-UV lamp (1000 W) in the range between 200-450 nm. Before UV radiation the PTFE pieces were dipped for 5 min into the pre-synthesized Au/IL dispersion (from thermal decomposition), then placed into quartz tubes without washing and irradiated.

Deposition of Au-NPs onto different carbon nanotubes (CNTs) e.g., SingleWall (SWCNTs) or MultiWall (MWCNTs) failed. Different possibilities were tested like, the thermal and photochemical decomposition of dissolved/dispersed Au(CO)Cl in the presence of dissolved CNTs (0.1-0.5 wt.%) in BMIm$^+$BF$_4^-$. Also post synthetic treatment at room temperature (RT) of dispersed Au-NPs mixed with dissolved CNTs (0.1-0.5 wt.%) in ILs failed. Further thermal treatment of these mixtures for several hours 2-6 h at (50-200°C) or photolytic irradiation for several minutes (5-20 min) was not successful.

The quantitative analysis of the gold content on the 10×10 mm^2 PTFE pieces (mass 0.017 g) cut from the sleeves was carried out by flame atomic absorbance spectroscopy (AAS) from a solution of 8.0 mL of conc. HCl/ HNO$_3$ (aqua regia) against an Au standard/calibration from Au(NO$_3$)$_3$. The AAS-determined gold contents on the PTFE pieces increased with the concentration of the gold dispersion as follows: 1) On untreated PTFE (blank) a blind value of 0.2855 mg/L or 0.0023 mg Au/PTFE piece was found and subsequently deducted from the measured Au values for the treated PTFE samples. 2) A 0.5 wt.% Au/IL dispersion gave 0.8235 mg/L or 0.0066 mg Au/PTFE piece. 3) A 1 wt.% Au/IL dispersion gave 9.1925 mg/L or 0.0735 mg/PTFE piece. 4) A 1.5 wt.% Au/IL gave 25.515 mg/L or 0.2041 mg Au/PTFE piece.

Growing of Au-NPs (Chapter 3.6)

Materials: $KAuCl_4$ and $SnCl_2$ (anhydrous) were obtained from Strem and Aldrich.

Au-NPs were obtained by the reduction of $KAuCl_4$ with $SnCl_2$ in different ILs $BMIm^+BF_4^-$, $BMIm^+PF_6^-$ or $BMIm^+OTf^-$. The salts were dissolved under argon atmosphere in the dried and deoxygenated ionic liquid. The growth process can be initiated and controlled by the slow dripping of a defined volume 0.15 – 1.2 mL (0.08 mmol $KAuCl_4$ dissolved in 3 g of different ILs) to a mixture of dissolved $SnCl_2$ (0.1055 mmol, excess, $SnCl_2$ dissolved in 2 g of different ILs) IL at room temperature under argon atmosphere and at a constant stirring rate (1000rpm).

During the Au growth process the colour and Au-NP size was controlled by the molar Au:Sn ratios (from 1:24, 1:12, 1:10, 1:8; 1:6, 1:5, 1:4.5, 1:4 to1:3) in $BMIm^+BF_4^-$ ($\delta = 1.208$ g/cm^3).

No.	V($KAuCl_4$)/mL [a]	ΔV / mL	n Au/mmol	~ molar ratio [b] n Au : n Sn
1	0.15 mL	0.15 mL	0.0048	1 : 24
2	0.3 mL	0.15 mL	0.0097	1 : 12
3	0.4 mL	0.1 mL	0.0129	1 : 10
4	0.5 mL	0.1 mL	0.0161	1 : 8
5	0.6 mL	0.1 mL	0.0193	1 : 6
6	0.7 mL	0.1 mL	0.0225	1 : 5
7	0.8 mL	0.1 mL	0.0258	1 : 4.5
8	0.9 mL	0.1 mL	0.0290	1 : 4
9	1.2 mL	0.3 mL	0.0387	1 : 3

[a] 30.0 mg (0.08 mmol) $KAuCl_4$ in 3.0 g (2.5 ml) IL, c(Au) = 0.03 mmol/mL from which a specified amount was added to 20.0 mg (0.106 mmol) $SnCl_2$ in 2.0 g (1.7 ml) $BMIm^+BF_4^-$. [b] The addition can be carried out step by step, starting with the 1:24 ratio (No.1), this is the addition of 0.15 mL of the $KAuCl_4$ solution with c(Au) = 0.03 mmol/mL. The next addition of 0.15 mL (No.2) results in the 1:12 ratio, and so on with further additions No.3-9.

REFERENCES

1. M. Hosokawa, K. Nogi, M. Naito, T. Yokoyama, *Nanoparticle Technology Handbook*, Elsevier, **2007**.
2. G. B. Sergeev, *Nanochemistry*, Elsevier B.V., Amsterdam, **2006**.
3. B. Bhushan, *Springer Handbook of Nanotechnology*, 2nd. Ed., Springer, **2007**.
4. C. N. R. Rao, A. Muller, A. K. Cheetham, *Chemistry of Nanomaterials*, Wiley-VCH, Weinheim, **2004**.
5. R. W. Seigel, H. Hu, M. C. Roco, *Nanostructure Science and Technology*, Kluwer Academic Publishers, Boston, **1999**.
6. Ch. Baerlocher, W.M. Meier, D.H. Olson, *Atlas of Zeolite Framework Types*, 5nd. Ed, Elsevier, **2001**.
7. R. P. Feynmann, *Miniaturization*, Reinhold, New York, **1961**.
8. M. C. Roco, R. S. William, A. P. Alivisatos, *Nanotechnology Research Directions*, Kluwer Academic Publishers, Boston, **2000**.
9. G. Schmid, *Nanoparticles - From Theory to Application*, Wiley-VCH, **2004**.
10. C. N. R. Rao, R. J. Thomas, G. U. Kulkarni, *Nanocrystals*, Springer, **2007**, pp. 5.
11. C. N. R. Rao and A. K. Cheetham, *J. Mater. Chem.* **2001**, *11*, 2887.
12. P. Chinni, *Gazz. Chim. Ital.* **1979**, *109*, 225; P. Chinni, *J. Organomet. Chem.* **1980**, *200*, 37.
13. J. Kepler, *Seu De Niue Sexangula*, Tampach, **1611**; George F. Szprio: *Kepler's Conjecture: How Some of the Greatest Minds in History Helped Solve One of the Oldest Math Problems in the World*, Wiley, **2003**.
14. G. A. Ozin, A. C. Arsenault, L. Cadermateri, *Nanochemistry*, 2nd ed.; RSC-Publishing, **2009**.
15. J. Lawrence, G. Xu, *Mat. Res. Soc. Symp. Proc.* **2002**, 704, 219.
16. D. Snow, C. Brumlik, Nanoparticles for Hydrogen Storage, Transportation and Distribution, U.S. Patent 6,589,312 B1, **2003**.
17. M.-I. Baraton, L. Merhari, J. Wang, and K. Gonsalves, *Nanotechnology* **1998**, *9*, 356.
18. J. Baran, Jr., and O. Cabrera, Use of Surface-Modified Nanoparticles for Oil Recovery, U.S. Publ., US 2003/0220204 A1, **2003**.
19. F. Caruso, Colloids and Colloid Assemblies, Wiley-VCH, Weinheim, **2004**, pp. 494.
20. A. Singhal, G. Skandan, F. Badway, A. Manthiram, H. Ye, J.J. Zhu, *J. Power Sources* **2004**, *129*, 38.
21. G.O. Mueller, and R. Mueller-Mach, Lumileds Lighting US, LLC, *Light-emitting devices utilizing nanoparticles*, European Patent Application, Appl. 03076772.7.
22. C.C. Koch, Mechanical and thermal processing methods, *Rev. Adv. Mater. Sci.* **2003**, *5*, 91.
23. Y. Todaka, M. Nakamura, S. Hattori, K. Tsuchiya, M. Umemoto, *Mat. Trans.* **2003**, *44*, 277.
24. Cabot and Degussa Technical, Literature, Personal communication, **2000**; www.apt-powders.com
25. Y. Gogotsi, *Handbook of Nanomaterials*, CRC Press, Taylor&Francis Group, **2006**, pp 20.
26. C. G. Granquist, and R.A. Buhrman, *J. Appl. Phys.* **1976**, *47*, 2200.
27. R. Birringer, H. Gleiter, H.P. Klein, and P. Marquardt, *Phys. Lett.* **1984**, 102A, 8, 365.
28. W. Chang, G. Skandan, H. Hahn, S.C. Danforth, and B.H. Kear, *Nanostruct. Mater.* **1994**, *4*, 345.
29. H. Gleiter, *Progress in Mat. Sci.* **1990**, *33*, 4.
30. M.T. Swihart, *Colloid Interface Sci.* **2003**, *8*, 127; V. Papaefthymiou, A. Kostikas, A. Simopoulos, J. Appl. Phys. **1990**, *67*, 4487.
31. G. Messing, S. Zhang, and V. Jayanthi, *J. Amer. Ceram. Soc.* **1993**, *76*, 2707.
32. Y. Gogotsi, *Handbook of Nanomaterials*, CRC Press, Taylor&Francis Group, **2006**, pp 22.
33. S. Tsai, Y. Song, Ch.Y. Chen, T. K. Tseng, C.S. Tsai, H.M. Lin, *Mat. Res. Symp. Proc.* **2002**, *704*, 85.
34. F.K. Urban III, A. Hosseini-Tehrani, P. Griffiths, *J. Vac. Sci. Technol.* **2002**, *B20*, 995.
35. R.C. Birtcher, S.E. Donnelly, S. Schlutig, *Phys. Rev. Lett.* **2000**, *85*, 4968.
36. S. Kim, and J. Maier, *J. Electrochem. Soc.* **2002**, 149, J73–J83.
37. V. Rotello, Nanoparticles – Building Blocks for Nanotechnology, Springer, **2004**.
38. C. J. Brinker, C. W. Scherer, *Sol–Gel Science*, Academic Press, San Diego, CA, **1990**.
39. L. V. Interrante, M. J. Hampden-Smith, *Chemistry of Adv. Materials*, Wiley-VCH, New York, **1998**.
40. M. D. Fokema, E. Chiu, J.Y. Ying, *Langmuir* **2000**, *16*, 3154.
41. V. T. Liveri, *Controlled Synthesis of Nanoparticles in Microheterogenous Systems*, Springer, **2006**.
42. Y. Gogotsi, *Handbook of Nanomaterials*, CRC Press, Taylor&Francis Group, **2006**, pp 24.
43. I. Capek, *Adv. Colloid Interface Sci.* **2004**, *110*, 49.
44. H. Huang, G.Q. Xu, W.S. Chin, L.M. Gan, and C.H. Chew, *Nanotechnology* **2002**, *13*, 318.
45. P.-Y. Silvert, R. Herrera-Urbina, N. Duvauchelle, *J. Mater. Chem.* **1996**, *6*, 573.
46. T. Miyao, T. Sawaura, S. Naito, *J. Mat. Sci. Lett.* **2002**, *21*, 867.

47 W.R. Moser, *Process for the preparations of Solid State Materials*, U.S. Patent 5,466,646, **1995**.
48 E. Redel, R. Thomann, C. Janiak, *Inorg. Chem.* **2008**, *48*, 14.
49 T. Gutel, J. Garcia-Antón, K. Pelzer, K. Philippot, C. C. Santini, L. S. Ott, R. G. Finke, *Inorg. Chem.* **2006**, *45*, 8382.
50 G. S. Fonseca, A. P. Umpierre, P. F. P. Fichtner, S. R. Teixeira, J. Dupont, *Chem. Eur. J.* **2003**, *9*, 3263.
51 Z. Li, A. Friedrich, A. Taubert, *J. Mater. Chem.* **2008**, *18*, 1008.
52 P. Migowski, G. Machado, L. M. Rossi, G. Machado, J. Morais, S. R. Teixeira, M. C. M. Alves, A. Traverse, J. Dupont, *Phys.Chem.Chem.Phys.* **2007**, *9*, 4814.
53 J. M. Zhu, Y. H. Shen, A. J. Xie, L. G. Qiu, Q. Zhang, X. Y. Zhang, *J. Phys. Chem. C* **2007**, *111*, 7629.
54 M.A. Firestone, M.L. Dietz, S. Seifert, S. Trasobares, D.J. Miller, N.J. Zaluzec, *Small* **2005**, *1*, 754.
55 K. Peppler, M. Polleth, S. Meiss, M. Rohnke, J.Z. Janek, *Z. Phys. Chem.* **2006**, *220*, 1507.
56 A. Safavi, N. Maleki, F. Tajabadi, E. Farjami, *Electrochem.Commun.* **2007**, *9*, 1963.
57 K. Kim, C. Lang, P.A. Kohl, *J. Electrochem. Soc.* **2005**, *152*, E9.
58 E. Redel, R. Thomann, C. Janiak, *Chem.Comm.* **2008**, 1789.
59 J. Krämer, E. Redel, R. Thomann, C. Janiak, *Organometallics* **2008**, *27*, 1976.
60 E. Redel, J. Krämer, R. Thomann, C. Janiak, *J. Organomet Chem.* **2009**, *694*, 1069.
61 D. O. Silva, J. D. Scholten, M. A. Gelesky, S. R. Teixeira, A. C. B. Dos Santos, E. F. Souza-Aguiar, J. Dupont, *ChemSusChem* **2008**, *1*, 291.
62 M. Scariot, D. O. Silva, J. D. Scholten, G. Machado, S. R. Teixeira, M. A. Novak, G. Ebeling, J. Dupont, *Angew. Chem. Int. Ed.* **2008**, *47*, 9075.
63 Fonseca, G. S.; Machado, G.; Teixeira, S. R.; Fecher, G. H.; Morais, J.; Alves, M. C. M.; Dupont, J. *J. Colloid Interface Sci.* **2006**, *301*, 193; Fonseca, G. S.; Domingos, J. B.; Nome, F.; Dupont, J. *J. Mol. Catal. A: Chem.* **2006**, *248*, 10; Fonseca, G. S.; Fonseca, A. P.; Teixeira, S. R.; Dupont, J. *Chem. Eur. J.* **2003**, *9*, 3263; Dupont, J.; Fonseca, G. S.; Umpierre, A. F.; Fichter, P. F. P.; Teixeira, S. R. *J. Am. Chem. Soc.* **2002**, *124*, 4228.
64 Zhu, J. M.; Shen, Y. H.; Xie, A. J.; Qiu, L. G.; Zhang, Q.; Zhang, X. Y. *J. Phys. Chem. C* **2007**, *111*, 7629; Firestone, M. A.; Dietz, M. L.; Seifert, S.; Trasobares, S.; Miller, D. J.; Zaluzec, N. J. *Small* **2005**, *1*, 754; Peppler, K.; Polleth, M.; Meiss, S.; Rohnke, M.; Janek, J. *Z. Phys. Chem.* **2006**, *220*, 1507; Safavi, A.; Maleki, N., Tajabadi, F.; Farjami, E. *Electrochem. Commun.* **2007**, *9*, 1963.
65 Silveira, E. T.; Umpierre, A. P.; Rossi, L. M.; Machado, G.; Morais, J.; Soares, G. V.; Baumvol, I. J. R.; Teixeira, S. R.; Fichtner, P. F. P.; Dupont, *J. Chem. Eur. J.* **2004**, *10*, 3734.
66 Gutel, T.; Garcia-Antón, J.; Pelzer, K.; Philippot, K.; Santini, C. C.; Chauvin, Y.; Chaudret, B.; Basset, J.-M. *J. Mater. Chem.* **2007**, *17*, 3290.
67 Migowski, P.; Machado, G.; Rossi, L. M.; Machado, G.; Morais, J.; Teixeira, S. R.; Alves, M. C. M.; Traverse, A.; Dupont, *J. Phys. Chem. Chem. Phys.* **2007**, *9*, 4814.
68 $Fe(CO)_5$ and $Ni(CO)_4$ are industrially produced on a multi-ton scale; see D. G. E. Kerfoot, Nickel, in Ullmann's Encyclopaedia of Industrial Chemistry, 5th ed. (online), Wiley, **2008**; E. Wildermuth, H. Stark, G. Friedrich, F. L. Ebenhöch, B. Kühborth, J. Silver, R. Rituper, Iron Compounds, in Ullmann's Encyclopaedia of Industrial Chemistry, 5th ed. (online), Wiley, **2008.**
69 G. J. Hutchings, M. Haruta, Appl. Catal. **2005**, *291*, 1–261; M. Haruta, *Nature* **2005**, *437*, 1098.
70 Rao, C. N. R.; Thomas, P. J.; Kulkarni, G.U. *Nanocrystals: Synthesis, Properties and Applications*, Springer-Verlag Berlin-Heidelberg, **2007**, 175 pages.
71 Barber, D. J.; Freestone, I. C. *Archaeometry*, **1990**, *32*, 33.
72 Zsigmondy, R. *Nobel Prize for Colloid Chemistry* **1925**.; Zsigmondy, R. *Z. Phys. Chem.* **1906**, *56*, 65.
73 J. Turkevich, P. C. Stevenson, J. Hillier, Discuss. Faraday Soc. **1951**, *11*, 55.
74 M. Faraday, Philos. Trans. *R. Soc. London* **1857**, *147*, 145; R. Zsigmondy, P. A. Thiessen, *Das Kolloide Gold*, Akademische Verlagsgesellschaft M.B.H., Leipzig, **1925**, pp. 22.
75 G. J. Hutchings, M. Brust, H. Schmidbaur, *Chem. Soc. Rev.* **2008**, 37, 1759.
76 G. Frens, Nature **1973**, *241*, 20.
77 Y. Mastai, G. Gedanken, Chemistry of Nanomaterials (C. N. R. Rao, A. Müller, A. K. Cheetham, eds.), Wiley-VCH, Weinheim **2004**, Vol. 1, pp. 212.
78 V. Myroshnychenko, J. Rodríguez-Fernández, I. Pastoriza-Santos, A. M. Funston, C. Novo, P. Mulvaney, L. M. Liz-Marzán, F. J. García de Abajo, *Chem. Soc. Rev.* **2009** in press.
79 J. A. Alonso, *Chem. Rev.* **2000**, *100*, 637.
80 C. Lee, Y. Kang, K. Lee, S. R. Kim, D.-J. Won, J. S. Noh, H. K. Shin, C. K. Song, Y. S. Kwon, H.-M. So, J. Kim, *Current Applied Physics* **2002**, *2*, 39.

81 P. Wasserscheid, *Ionic Liquids in Synthesis*, 2en. Edition, Vol.1-2, Wiley-VCH, **2007**.
82 W. Hückel, *Chem. Ber.* **1958**, *91*, XIX–LXVI.
83 P. Walden, *Bull. Acad. Impér. Sci. St. Pétersbourg* **1914**, *8*, 405.
84 A. Stark and K. R. Seddon, in *'Kirk-Othmer Encyclopaedia of Chemical Technology'*, ed. A. Seidel, John Wiley & Sons, Inc., Hoboken, New Jersey, **2007**, vol. 26, pp. 836.
85 M. Freemantle, *Chem. Eng. News* **2004**, *82*, 10.
86 G. W. Meindersma, M. Maase, A. B. De Haan, *Ullmans Electronic Edition 5th*, Wiley-VCH, **2007**.
87 D. J. Adams, P. J. Dyson, S. J. Tavener, *Chemistry in Alternative Reaction Media*, Wiley-VCH, Weinheim **2004**, pp 75.
88 D. Astruc, *Nanoparticle in Catalysis*, Wiley-VCH, Weinheim, **2008**.
89 J. D. Holbrey, K. R. Seddon, *Clean Products and Processess,* **1999**, *1*, 223.
90 R. D. Rogers and K. R. Seddon, Science **2003**, *302*, 792; M. Maase, in *'Multiphase Homogeneous Catalysis'*, ed. B. Cornils, W. A. Herrmann, I. T. Horvath, W. Leitner, S. Mecking, H. Olivier-Bourbigou and D. Vogt, Wiley-VCH, Weinheim, **2005**, vol. 2, pp. 560.
91 The BASIL process", http://www2.basf.de/en/intermed/nbd/products/ionic_liquids/processes/acid.htm 2005
92 http://www.corporate.basf.com/de/innovationen/2006/jetstream.htm
93 http://www.corporate.basf.com/de/ innovationen/preis/2004/basil.htm
94 M. Maase, K. Massonne, K. Halbritter, R. Noe, M. Bartsch, W. Siegel, V. Stegmann, M. Flores, O. Huttenloch and M. Becker, *Method for the separation of acids from chemical reaction mixtures by means of ionic fluids*, World Pat., WO 2003 062171, **2003**.
95 J. Baker, ECN Innovation Awards 2004 – the winners!, *Eur.Chem. News*, **2004**, 18.
96 Y. Chauvin, J. F. Gaillard, D. V. Quang and J. W. Andrews, *Chem. Ind.* **1974**, 375.
97 Axens, "Dimersol-X", http://www.axens.net/html-gb/offer/offer_processes_70.html.php (**2007**).
98 H.-G. Elias, Makromoleküle, Vol. 1-4, 6nd. Ed., Wiley-VCH, Weinheim, **2001**.
99 Y. Chauvin, A. Hirschauer and H. Olivier, *J. Mol. Catal.* **1994**, *92*, 155; Y. Chauvin, S. Einloft and H. Olivier, *Ind. Eng. Chem. Res.* **1995**, *34*, 1149.
100 F. Favre, A. Forestie`re, F. Hugues, H. Olivier-Bourbigou and J. A. Chodorge, Oil Gas-Eur. Mag., **2005**, *31*, 83; Y. Chauvin, B. Gilbert and I. Guibard, *Chem. Commun.,* **1990**, 1715.
101 R. D. Rogers, *Green Chem.* **2004**, *6*, G17–G19.
102 BASF, "Cellulose processing", http://www2.basf.de/en/intermed/ nbd/products/ionic_liquids/applications/cellulose/?id=mGwEvAf1Ebw23hM; M. Maase and V. Stegmann, Solubility of cellulose in ionic liquids with addition of amino bases, World Pat., WO 2006 108861 (**2006**); M. Maase and V. Stegmann, Solutions of cellulose in ionic liquids, DE Pat., 102005017715 (**2006**).
103 F. Endres, ChemPhysChem, **2002**, *3*, 144.
104 F. Endres, *Phys. Chem. Chem. Phys.* **2006**, *8*, 2101.
105 P. K. Lai, M. Skyllas-Kazacos, J. Electroanal. Chem., 1988, 248, 431;Y. Zhao, T. J. VanderNoot, *Electrochim. Acta* **1997**, *42*, 3; M. R. Ali, A. Nishikata, T. Tsuru, Indian *J. Chem. Technol.* **1999**, *6*, 317.
106 El Abedin, E. M. Moustafa, R. Hempelmann, H. Natter, F. Endres, *Electrochem. Commun.* **2005**, *7*, 1111; S. Z. El Abedin and F. Endres, *ChemPhysChem*, **2006**, *7*, 58; F. Endres, S. Z. El Abedin and N. Borissenko, *Z. Phys. Chem.* **2006**, *220*, 1377; Q. X. Liu, S. Z. El Abedin and F. Endres, Surf.Coat. Technol. **2006**, *201*, 1352.
107 IoLiTec web.site, homepage / www.iolitec.de or www.iolitec.com
108 N. V. Plechkova, K. R. Seddon, Chem.Soc.Rev., **2008**, *37*, 123.
109 M. J. Earle, J. M. S. S. Esperanca, M. A. Gilea, J. N. C. Lopes, L. P. N. Rebelo, J. W. Magee, K. R. Seddon, J. A. Widegren, Nature **2006**, *439*, 831.
110 J. P. Leal, J. M. S. S. Esperanc‚a, M. E. M. da Piedade, J. N. C. Lopes, L. P. N. Rebelo and K. R. Seddon, *J. Phys. Chem. A*, **2007**, *111*, 6176.
111 Sheldon, R.; *Chem. Commun.* **2001**, 2399; Wasserscheid, P.; Keim, W.; *Angew. Chem. Int. Ed.* **2000**, *39*, 3773; Dupont, J.; Consorti, C. S.; Spencer, J.; *J. Braz. Chem.Soc.* **2000**, *11*, 337; Welton, T.; *Chem. Rev.* **1999**, *99* 2071;
112 F. de Souza, R. F.; Padilha, J. C.; Gonçalves, R. S.; Dupont, J.;*Electrochem. Comm.* **2003**, *5*, 728.
113 Wang, P.; Zakeeruddin, S. M.; Comte, P.; Exnar, I.; Gratzel, M.; *J . Am. Chem. Soc.* **2003**, *125*, 1166.
114 Reich, R. A.; Stewart, P. A.; Bohaychick, J.; Urbanski, J. A.; *Lub. Eng.* **2003**, *59*, 16; Ye, C. F.; Liu, W. M.; Chen, Y. X.; Yu, L. G.; *Chem. Commun.* **2001**, 2244.

115 Anderson, J. L.; Armstrong, D. W.; *Anal. Chem.* **2003**, *75,* 4851; Armstrong, D. W.; He, L. F.; Liu, Y. S.; *Anal. Chem.* **1999**, *71,* 3873.
116 Armstrong, D. W.; Zhang, L-K.; He, L.; Gross, M. L.; *Anal. Chem.* **2001**, *73,* 3679-3686; Dyson, P. J.; McIndoe, J. S.; Zhao, D. B.; *Chem. Commun.* **2003**, 508;
117 Lozano, P.; de Diego, T.; Carrie, D.; Vaultier, M.; Iborra, J. L.; *Biotech. Prog.* **2003**, *19,* 380.
118 Branco, L. C.; Crespo, J. G.; Afonso, C. A. M.; *Angew. Chem. Int. Ed.* **2002**, *41,* 2771; Branco, L. C.; Crespo, J. G.; Afonso, C. A. M.; *Chem. Eur. J.* **2002**, *8,* 3865.
119 Holbrey, J. D.; Seddon, K. R.; *J. Chem. Soc. Dalton Trans.* **1999**, 2133.
120 Adams, C. J.; Bradley, A. E.; Seddon, K. R.; *Aust. J. Chem.* **2001**, *54,* 679.
121 Dupont, J.; Fonseca, G. S.; Umpierre, A. P.; Fichtner, P. F. P.; Teixeira, S. R.; *J. Am. Chem. Soc.* **2002**, *124,* 4228; Fonseca, G. S.; Umpierre, A. P.; Fichtner, P. F. P.; Teixeira, S. R.; Dupont, J.; *Chem. Eur. J.* **2003**, *9,*3263.
122 Carmichael, A. J.; Hardacre, C.; Holbrey, J. D.; Nieuwenhuyzen, M.; Seddon, K. R.; *Mol. Phys.* **2001**, *99,* 795.
123 Majewski, P.; Pernak, A.; Grzymislawski, M.; Iwanik, K.; Pernak, J.; *Acta Histochemi,* **2003**, *105,* 135.
124 Welton, T.; *Chem. Rev.* **1999**, *99* 2071; Seddon, K. R.; *J. Chem. Technol. Biotechnol.* **1997**, *68,* 351; Chauvin, Y.; *Act. Chim.* **1996**, 44.
125 Rogers, R. D.; Seddon, K. R.; *Ionic Liquids Industrial Applications to Green Chemistry* **2001**, ACS, Symposium Series 818.
126 Data obtained from Cambridge Crystallographic Data Center (www. ccdc.can.ac.uk.).
127 J. Dupont, J. Braz. Chem. Soc. **2004**, *15,* 341.
128 Van den Broeke, J.; Stam, M.; Lutz, M.; Kooijman, H.; Spek, A. L.; Deelman, B. J.; van Koten, G.; *Eur .J .Inorg. Chem.* **2003**, 2798.
129 Amyes, T. L.; Diver, S. T.; Richard, J. P.; Rivas, F. M.; Toth, K.; *J. Am. Chem. Soc.* **2004**, in press.
130 For a detailed discussion concerning the different types of hydrogen bonds see: Jeffrey, G. A.; *An Introduction to Hydrogen Bonding*, Oxford University Press: Oxford, **1997**.
131 Antonietti, M., Kuang, D.B., Smarsly, B. and Yong, Z., *Angew. Chem. Int. Ed.*, **2004**, 43, 4988.
132 Wasserscheid, P. Welton, T., eds.; *Ionic Liquids in Synthesis*, 2nd, Wiley-VCH: New York, **2007**
133 Holbrey, J. D.; Reichert, W. M.; Nieuwenhuyzen, M.; Sheppard, O.; Hardacre, C.; Rogers, R. D.; *Chem. Commun.* **2003**, 476.
134 Bronger, R. P. J.; Silva, S. M.; Kamer, P. C. J.; van Leeuwen, P. W. N. M.; *Chem. Commun.* **2002**, 3044.
135 Lozano, P.; de Diego, T.; Carrie, D.; Vaultier, M.; Iborra, J. L.; *Biotech. Prog.* **2003**, *19,* 380.
136 Swatloski, R. P.; Spear, S. K.; Holbrey, J. D.; Rogers, R. D.; *J. Am. Chem. Soc.* **2002**, *124,* 4974.
137 Huang, J.; Jiang, T.; Han, B. X.; Gao, H. X.; Chang, Y. H.; Zhao, G. Y.; Wu, W. Z.; *Chem.Commun.* **2003**, 1654; Dupont, J.; Fonseca, G. S.; Umpierre, A. P.; Fichtner, P. F. P.; Teixeira, S. R.; *J. Am. Chem. Soc.* **2002**, *124,* 4228.
138 Redel, E.; Thomann, R.; Janiak, C. *Inorg. Chem.* **2008**, *47,* 14; Redel, E.; Thomann, R.; Janiak, C. *Chem.Commun.* **2008**, 1789.
139 Verwey, E.J.W. and Overbeek, J.T.G. (Ed.) *Theory of the Stability of Lyophobic Colloids*, Dover Publications, Inc., New York, **1999**.
140 Bostrom, M., Williams, D.R.M. and Ninham, B.W. *Phys. Rev. Lett.*, **2001**, 87, Article No. 168103.
141 Ott, L.S., Cline, M.L., Deetlefs, M., Seddon, K.R. and Finke, R.G., *J. Am. Chem. Soc.* **2005**, *127,* 5758.
142 A. P. Umpierre, G. Machado, G. H. Fechner, J. Morias, J. Dupont, *Adv. Synth. Catal.*, **2005**, *347,* 1404.
143 G. Machado, J.D. Scholten, T. de Vargas, S. R: Teixeira, L. H. Ronchi, J. Dupont, *Int. J. Nanotechnol.*, **2007**, *4,* 541.
144 R. Elghanian, J.J. Storhoff, R.C. Mucic, R.L. Letsinger, C.A. Mirkin, Science **1997**, *277,* 1078.
145 J.A. Alonso, Chem. Rev. **2000**, *100,* 637.
146 V. I. Pârvulescu and C. Hardacre, *Chem. Rev.*, **2007**, *107,* 2665.
147 Esteban-Cubillo, A.; Díaz, C.; Fernández, A.; Díaz, L. A.; Pecharromán, C.; Torrecillas, R.; Moya, J. S. *J. Eur. Ceram. Soc.* **2006**, *26,* 1; Hodnet, B. K. *Heterogeneous Catalytic Oxidation*, 1 ed,; Wiley-VCH, Weinheim, **2000**, pp. 163.
148 Wang, R.; Guo, X.; Wang, X.; Hao, J. *Catal. Lett.* **2003**, *90,* 57.
149 Braunstein, P.; Rosé, J. *Metal Clusters in Chemistry* (Braunstein, P.; Oro, L. A.; Raithby, P. R.; eds.) Wiley-VCH, Weinheim, **2001**, Vol. 2, ch. 2, pp. 616.
150 A.H. Lu, E.L. Salabas, F. Schüth, Angew. Chem. Int. Ed. **2007**, *46,* 1222.
151 A. Gedanken, Ultrasonics Sonochemistry **2004**, *11,* 47.

152 C.N.R. Rao, S.R.C. Vivekchand, K. Biwas, A. Govindaraj, Dalton Trans. **2007**, 3728.
153 Y. Mastai, A. Gedanken, Chemistry of Nanomaterials (C.N.R. Rao, A. Müller, A.K. Cheetham, eds.), Wiley-VCH, Weinheim **2004**, Vol. 1, pp. 113.
154 D. Mahajan, E.T. Papish, K. Pandya, Ultrasonics Sonochemistry **2004**, *11*, 385.
155 J. Park, J. Joo, S.G. Kwon, Y. Jang, T. Hyeon, Angew. Chem. Int. Ed. **2007**, *46*, 4630.
156 Ott, L. S.; Finke, R. G. *Inorg. Chem.* **2006**, *45*, 8382; Besson, C.; Finney, E. E; Finke, R. G. *J. Am. Chem. Soc.* **2005**, *127*, 8179.
157 Ott, L. S.; Cline, M. L.; Deetlefs, M.; Seddon, K. R.; Finke, R. G. *J. Am. Chem. Soc.* **2005**, *127*, 5758.
158 Dupont, J. *J. Brazil Chem. Soc.* **2004**, *15*, 341; Consorti, C. S.; Suarez, P. A. Z.; de Souza, R. F.; Burrow, R. A.; Farrar, D. H.; Lough, A. J.; Loh, W.; da Silva, L. H. M.; Dupont, J. *J. Phys. Chem. B* **2005**, *109*, 4341; Dupont, J.; Suarez, P. A. Z.; de Souza, R. F.; Burrow, R. A.; Kintzinger, J. P. *Chem. Eur. J.* **2000**, *6*, 2377.
159 Verwey, E. J. W.; Overbeek, J. T. G. *Theory of the Stability of Lyo-hobic Colloids*, 2nd ed.; Dover Publications: Mineola, New York, 1999; Ninham, B. W. *Adv. Coll. Int. Sci.* **1999**, *83*, 1.
160 D.; Lu, F.; Aranzes, J. R. *Angew. Chem. Int. Ed.* **2005**, *44*, 7852; Antonietti, M.; Kuang, D.; Smarly, B.; Zhou, Y. *Angew. Chem. Int. Ed.* **2004**, *116*, 5096.;
161 A. Taubert, Z. Li, Dalton Trans. **2007**, 723.
162 J. Dupont, J. Brazil. Chem. Soc. **2004**, 15, 341.
163 C.S. Consorti, P.A.Z. Suarez, R.F. de Souza, R.A. Burrow, D.H. Farrar, A.J. Lough, W. Loh, L.H.M. da Silva, J. Dupont, J. Phys. Chem. B **2005**, 109, 4341.
164 J. Dupont, P.A.Z. Suarez, R.F. de Souza, R.A. Burrow, J.P. Kintzinger, Chem. Eur. J. **2000**, 6, 2377
165 E.J.W. Verwey, J.T.G. Overbeek, Theory of the Stability of Lyophobic Colloids, 2 ed., Dover Publications Mineola, New York, **1999**.
166 B.W. Ninham, *Adv. Coll. Int. Sci.* **1999**, 1, 83.
167 L.S. Ott, R.G. Finke, *Coord. Chem. Rev.* **2007**, 251, 1075.
168 A.N. Shipway, E. Katz, I. Willner, *ChemPhysChem* **2000**, 1, 18.
169 T. Cassagneau, J.H. Fendler, *J. Phys. Chem. B.* **1999**, 103, 1789.
170 C.D. Keating, K.K. Kovaleski, M.J. Natan, *J. Phys. Chem. B.* **1998**, 102, 9404.
171 E. Redel, J. Krämer, R. Thomann, C. Janiak, *GIT Labor-Fachzeitschrift* **2008**, 04, 400.
172 Schubert, T. J. S. *Nachr. Chem.* **2005**, *53*, 1222;
173 STOE WinXPow version 1.10, data base, STOE & Cie GmbH, Darmstadt, Germany, 2002.
174 Wang, Y.; Maksimuk, S.; Shen R.; Yang, H. *Green Chem.* **2007**, *9*, 1051.
175 A.F. Holleman, N. Wiberg, Lehrbuch der Anorganischen Chemie, 102 ed., Walter de Gruyter, Germany, **2007** Berlin, pp.1782.
176 R. Allmann, Röntgenpulverdiffraktometrie, Springer, Berlin, **2005**.
177 M. Scariot, D. O. Silva, J. D. Scholten, G. Machado, S. R. Teixeira, M. A. Novak, G. Ebeling, J. Dupont, Angew. Chem. Int. Ed. **2008**, *47*, 9075.
178 I. Krossing, J.M. Slattery, Z. Phys. Chem. **2006**, 220, 1343.
179 A. F. Holleman, N. Wiberg, *Lehrbuch der Anorganischen Chemie*, 102nd ed., Walter de Gruyter, Berlin, **2007**, pp. 1481–1482.
180 J. D. Scholten, G. Ebeling, J. Dupont, *Dalton Trans.* **2007**, 5554–5560; D. Astruc, *Nanoparticles and Catalysis*, Wiley, New York, **2007**, pp. 22–24; V. Calo, A. Nacci, A. Monopoli, S. Laera, N. Coffi, *J. Org. Chem.* **2003**, *68*, 2929; R. R. Deshmuhk, R. Rajagopal, K. V. Srinivasan, *Chem. Commun.* **2001**, 1544; L.Xu, W. Chen, J. Xiao, *Organometallics* **2000**, *19*, 1123.
181 P. J. Barnard, M. V. Baker, S. J. Berners-Price, B. W. Skelton, A. H. White, *Dalton Trans.* **2004**, 1038-1047; V. J. Catalano, A. L. Moore, *Inorg. Chem.* **2005**, *44*, 6558-6566;V. J. Catalano, A. O. Etogo, *Inorg. Chem.* **2007**, *46*, 5608-5615.
182 P. Dash, R. W. J. Scott, Chem. Commun. **2009**, DOI: 10.1039/b816446k.
183 H. Häkkinen, Chem. Soc. Rev. **2008**, *37*, 1847.
184 J. P. Tierney, P. Lidström, Microwave Assisted Organic Synthesis, Blackwell Publishing, **2005**, pp. 20 and pp. 140; C. J. Coleman, J. Austral. Math. Soc. Ser. B, **1991**, *33*, 1.
185 I. Krossing, J. M. Slattery, Z. Phys. Chem. **2006**, *220*, 1343.
186 A. N. Shipway, E. Katz, I. Willner, ChemPhysChem **2000**, *1*, 18.
187 T. Cassagneau, J. H. Fendler, J. Phys. Chem. B. **1999**, *103*, 1789.
188 C. D. Keating, K. K. Kovaleski, M. J. Natan, *J. Phys. Chem. B.* **1998**, *102*, 9404.
189 E. Redel, J. Krämer, R. Thomann, C. Janiak, *GIT Labor-Fachzeitschrift* **2008**, *04*, 400.

190 T. J. Gannon, G. Law, R. P. Watson, A. J. Carmichael, K. R. Seddon, *Langmuir* **1999**, *15*, 8429.
191 G. Law, R. P. Watson, A. J. Carmichael, K. R. Seddon, *Phys. Chem. Chem. Phys.* **2001**, *3*, 2879.
192 J. N. C. Lopes, M. F. C. Gomes, A. A. H. Pádua, *J. Phys. Chem. B.* **2006**, *110*, 16816.
193 J. N. C. Lopes, A. A. H. Pádua, *J. Phys. Chem. B.* **2006**, *110*, 3330.
194 C. S. Weisbecker, M. V. Merritt, G. M. Whitesides, *Langmuir* **1996**, *12*, 3763.
195 S. Chen, K. Kimura, *Chem. Lett.* **1999**, 1169.
196 STOE WinXPow version 1.10, data base, STOE & Cie GmbH, Darmstadt, Germany, 2002.
197 O. P. Khatri, K. Adachi, K. Murase, K-I. Okazaki, T. Torimoto, N. Tanaka, S. Kuwabata, H. Sugimura, *Langmuir*, **2008**, *24*, 7785.
198 Y. Shichibu, Y. Negishi, T. Tsukuda, T. Teranishi, *J. Am. Chem. Soc.* **2005**, *127*, 13464.
199 R. Balasubramanian, R. Guo, A. J. Mills, Royce W. Murray, *J. Am. Chem. Soc.* **2005**, *127*, 8126.
200 M. Walter, J. Akola, O. Lopez-Acevedo, P. D. Jadzinsky, G. Calero, C. J. Ackerson, R. L Whetten, H. Grönbeck, H. Häkkinen, *PNAS* **2008**, *105*, 9157.
201 J. Akola, M. Walter, R. L. Whetten, H. Häkkinen, H. Grönbeck, *J. Am. Chem. Soc.* **2008**, *130*, 3756.
202 P. Dutta, S. Pal, M. S. Seehra, M. Anand, C. B. Roberts, *Appl. Phys. Lett.* **2007**, *90*, 213102.
203 A. P. Umpierre, G. Machado, G. H. Fechner, J. Morais, J. Dupont, *Adv. Synth. Catal.* **2005**, *347*, 1404.
204 G. Machado, J. D. Scholten, T. de Vargas, S. R. Teixeira, L. H. Ronchi, J. Dupont, *Int. J. Nanotechnol.* **2007**, *4*, 541.
205 S. Özkar, R. G. Finke, *J. Am. Chem. Soc.* **2002**, *124*, 5796.
206 Z. Wang, Q. Zhang, D. Kuehner, A. Ivasaka, L. Niu, *Green Chemistry* **2008**, *10*, 907; Ki-Sub Kim, D. Demberelnyamba, H. Lee, *Langmuir* **2004**, *20*, 556.
207 Consorti, C. S.; Suarez, P. A. Z.; de Souza, R. F.; Burrow, R. A.; Farrar, D. H.; Lough, A. J.; Loh, W.; da Silva, L. H. M.; Dupont, J. *J. Phys. Chem. B* **2005**, *109*, 4341.
208 Verwey, E. J. W.; Overbeek, J. T. G. *Theory of the Stability of Lyop-hobic Colloids*, 2nd ed.; Dover Publications: Mineola, New York, 1999; Ninham, B. W. *Adv. Coll. Int. Sci.* **1999**, *83*, 1.
209 Ott, L. S.; Finke, R. G. *Coord. Chem. Rev.*, **2007**, *251*, 1075.
210 G. A. Ozin, A. C. Arsenault, L. Cadermateri, *Nanochemistry*, 2nd ed.; RSC-Publishing, **2009**; Astruc, D.; Lu, F.; Aranzes, J. R. *Angew. Chem. Int. Ed.* **2005**, *44*, 7852.
211 Abécassis, B.; Testard, F.; Spalla, O.; Barboux, P. *Nano Letters*, **2007**, *7*, 1723.
212 Becker, J.; Schubert, O.; Sönnichsen C. *Nano Letters*, **2007**, *7*, 1664.
213 V. K. La Mer, R. H. Dinegar, Angew. Chem. Int. Ed. Engl. **2004**, *43*, 6042; C. B. Murray, C. R. Kagan, M. G. Bawendi, *Annu. Rev. Mater. Sci.* **2000**, *30*, 545.
214 C. Sönnichsen, W. Fritsche, *100 years of nanoscience with the ultramicroscope*, Shaker Verlag, Aachen **2007**, pp. 100
215 I. Hussain, S. Graham, Z. Wang, B. Tan, D. C. Sherrington, S.P. Rannard, A.I. Cooper, M. Brust, *J. Amer. Chem. Soc.* **2005**, *127*, 16398.
216 J. Blackman, *Metallic Nanoparticles*, Elsevier, the Netherlands, Amsterdam, **2009**, pp. 113.
217 F. Endres, S. Z. El Abedin, *Phys. Chem. Chem. Phys.* **2006**, *8*, 2101.
218 P. Wasserscheid, T. Welton, *Ionic Liquids in Synthesis*, 2nd ed., Wiley-VCH, Weinheim, **2007**, 32.
219 R. P. Swatloski, J. D. Holbrey, R. D. Rogers, *Green Chemisty* **2003**, *5*, 36.
220 G. A. Baker, S. N. Baker, *Aust. J. Chem.* **2005**, *58*, 174.

Appendix: Organic-Inorganic Hybrid Materials

In the beginning of my Ph.D. research the project of halometallates, in particular halocuprates was followed besides the investigation of the nanoparticle area. With the results in the field of metal nanoparticle (M-NP) synthesis in ionic liquids (ILs) it was decided to leave the halometallate research and to focus solely on the M-NP/IL field. This appendix summarizes the results from the halocuprate research.

Networks with *inorganic bridging ligands* and *terminal organic ligands or* just *organic counterions* should better be called inorganic-organic hybrid materials. An important subgroup in inorganic-organic hybrid materials are halometallates, $[M_xX_y]^{n-}$ [1,2] which are of continuous interest for their structural richness, semiconducting,[3] magnetic,[4] photo-luminescent and photo-chromic properties.[5] These halometallates $[M_xX_y]^{n-}$ typically have organic counter-cations derived from alkyl or aromatic amine molecules by protonation. The ammonium cations do not coordinate to the metal atoms but serve as template molecules in the lattice of the 0D, 1D or 2D $[M_xX_y]^{n-}$ motif.

Organic-inorganic hybrid materials represent the natural interface between two worlds of chemistry (organic and inorganic) each with very significant contributions to the field of materials science. Through the synthesis of hybrid materials chemists want to take advantages of the best properties of each component (organic/inorganic) that forms the structure. In the hybrid approach there is an immense variety of adducts that can be formed between organic and inorganic species. Taking into account the great diversity of extended and molecular inorganic species (e.g. MX_n^-) and small organic molecules (e.g. NR_4^+) for the design of these hybrid materials, it is clear that the variety of combinations is tremendously high.

In our studies we focused our research on 1-dimensional halocuprates(I), which have been synthesized under hydrothermal conditions by Cu(II)/Cu(0) comproportionation and also on inclusion (host-guest) polyiodide species, I_n^{m-}.

We report the synthesis and physical investigation of the compound $\{(C_4H_{12}N_2)_2[Cu^I I_4](I_2)\}_n$ which agrees in its dark-blue crystal color, its mixture of I^-, I^{3-} and linear I_4^{2-} or linear I^{5-} polyiodide species in a linear channel arrangement, its channel diameter of ~5.5 Å and in the helical arrangement of the hydrogen bonded $\{(C_4H_{12}N_2)_2[Cu^I I_4]\}^+$ supramolecular host around the channels with the description of the classical, yet structurally elusive, starch-iodine compound.

Furthermore, 1-dimensional halocuprate(I) chains $[(Cu_2X_4)^{2-}]_n$ (= $[(CuX_2)^-]_n$, X = Cl, Br, I) have been synthesized under hydrothermal conditions through in-situ reduction of Cu(II)X_2 with Fe(II)X_2 or as phase pure materials through comproportionation of Cu(II)X_2 or Cu(II)O with Cu(0) metal in the presence of the respective aqueous hydrogen halide HX and a templating amine. Chains of *trans* edge-sharing tetrahedra are obtained with piperazinium or ethylenediammonium dications, while the 4,4'-bipyridinium dication gave chains of *cis* edge-sharing tetrahedra. Two monoprotonated piperazinium groups act as cationic ligands (Hpipz$^+$) towards copper atoms in a molecular [Cu$_4$(μ-Br$_6$)(Hpipz)$_2$] cluster. Electrical crystal conductivities of the halocuprate $[(Cu_2X_4)^{2-}]_n$ (= $[(CuX_2)^-]_n$) chains (X = Cl, Br, I) are around 10^{-8} S/cm at room temperature.

1 N. Louvain, N. Mercier and M. Kurmoo, *Eur. J. Inorg. Chem.* **2008**, 1654; M. A. Tershansy, A. M. Goforth, M. D. Smith and H.-C. zur Loye, *J. Chem. Cryst.* **2007**, *38*, 453; C.-L. Chen and A. M. Beatty, *Chem. Commun.*, 2007, 76; N. Mercier, *CrystEngComm* 2005, *7*, 429; M. C. Aragoni, M. Arca, C. Caltagirone, F. A. Devillanova, F. Demartin, A. Garau, F. Isaia and V. Lippolis, *CrystEngComm* **2005**, *7*, 544; A. M. Goforth, L. Peterson, Jr., M. D. Smith and H.-C. zur Loye, *J. Solid State Chem.* **2005**, *178*, 3553; N. Mercier and A. Riou, *Chem. Commun.* **2004**, *844*; U. Bentrup, M. Feist and E. Kemnitz, *Prog. Solid State Chem.* **1999**, *27*, 75.
2 Halocuprates: E. Redel, C. Röhr and C. Janiak, *Chem. Commun.* **2009**, DOI: 10.1039/b820151j; E. Redel, M. Fiederle and C. Janiak, *Z. Anorg. Allg. Chem.* **2009**, DOI: 10.1002/zaac.200900091; C. H. Arnby, S. Jagner, I. Dance, *CrystEngComm*, **2004**, *6*, 257; S. Jagner and G. Helgesson, *Adv. Inorg. Chem.*, 1991, *37*, 1; L. Subramanian and R. Hoffmann, *Inorg. Chem.* **1992**, *31*, 1021.
3 H.-H. Li, L.-G. Sun, Z.-R. Chen, Y.-J. Wang and J.-Q. Li, *Aust. J. Chem.* **2008**, *61*, 391-396; M. Okubo, M. Enomoto and N. Kojima, *Synth. Metals* **2005**, *125*, 461; D. B. Mitzi, *J. Mater. Chem.* **2004**, *14*, 2355.
4 P. Rabu and M. Drillon, *Adv. Eng. Mater.* **2003**, *5*, 189; S. Flandrois, N. B. Chanh, R. Duplessix, T. Maris and P. Négrier, *Phys. Status Solidi A* **1995**, *149*, 697.
5 N. Mercier, S. Poiroux, A. Riou and P. Batail, *Inorg. Chem.* **2004**, *43*, 8361; T. Gebauer and G. Schmid, *Z. Anorg. Allg. Chem.* **1999**, *625*, 1124; D. B. Mitzi and K. Liang, *Chem. Mater.* **1997**, *9*, 2990; T. Ishihara, J. Takahashi and T. Goto, *Solid State Commun.* **1989**, *69*, 933.

Die VDM Verlagsservicegesellschaft sucht für wissenschaftliche Verlage abgeschlossene und herausragende

Dissertationen, Habilitationen, Diplomarbeiten, Master Theses, Magisterarbeiten usw.

für die kostenlose Publikation als Fachbuch.

Sie verfügen über eine Arbeit, die hohen inhaltlichen und formalen Ansprüchen genügt, und haben Interesse an einer honorarvergüteten Publikation?

Dann senden Sie bitte erste Informationen über sich und Ihre Arbeit per Email an *info@vdm-vsg.de*.

Sie erhalten kurzfristig unser Feedback!

VDM Verlagsservicegesellschaft mbH
Dudweiler Landstr. 99
D - 66123 Saarbrücken

Telefon +49 681 3720 174
Fax +49 681 3720 1749

www.vdm-vsg.de

Die VDM Verlagsservicegesellschaft mbH vertritt

Printed by Books on Demand GmbH, Norderstedt / Germany